ST/SPACE/5

Office for Outer Space Affairs
United Nations Office at Vienna

SEMINARS

of the
United Nations Programme on
Space Applications

Selected Papers from Activities Held in 2000

UNITED NATIONS
New York, 2001

INTRODUCTION

This publication contains selected papers from workshops organized by the United Nations Programme on Space Applications in 2000.

The many practical benefits from space technology will play a central role in international development efforts. New applications of space technology are constantly being developed in diverse fields such as telecommunications, land use management, and health care, many taking advantage of the unique, broad perspective of satellites, which makes them efficient over large areas with little ground infrastructure. The fast-moving field of space technology is one of the more difficult for scientists in developing countries to enter, but is one in which a relatively small number of experienced professionals can make an enormous contribution to a country or a region. For this reason, the United Nations established the Programme on Space Applications as one of the priority activities for the Office for Outer Space Affairs. The Programme organizes eight to ten seminars and training courses annually for students and professionals from developing countries, and aims to increase the skill with which space technology is used in the development process worldwide.

This publication is the twelfth in an annual series beginning in 1989. It is part of the Office's efforts to encourage international exchange of experiences on practical applications of space technology. The papers published here are chosen among many presented at workshops of the Programme on Space Applications. In particular, papers are chosen that discuss policy issues and practical applications of space technology, since a number of scientific and technical journals exist in which participants of the United Nations workshops can and do publish research work. Workshops of the Programme on Space Applications are held throughout the world; papers discuss issues in a variety of countries and regions and papers are published in the language of submission.

CONTENTS

UNISPACE III: RECOMMENDATIONS FOR THE SPACE MILLENNIUM

United Nations Office for Outer Space Affairs

I. INTRODUCTION

As we stand at the turn of the millennium, we realize that there are many things to deplore and to correct. Grinding poverty and striking inequality persist within and among countries even amidst unprecedented wealth. Nature's life-sustaining capabilities are being seriously challenged by our own everyday activities. Despite spectacular advances in science and technology in recent years, more than a quarter of the world's population still remain in severe poverty. About three quarters of the world's poorest, who depend on agricultural activities for their living, continue to be caught in a downward spiral of poverty and degenerating environment, being forced to deplete resources to survive and then further impoverished by degradation of the environment.

Globalization has increased interdependence of the world's people and calls for shared values and a shared commitment to the human development of all people. Globalization of financial activities and global competition in the free markets challenge economic security and job security. Global transmission of pathogens threatens health security. Cultural identity and security are also being challenged by global broadcasting networks. Transnational organized crime threatens personal security. Transborder pollution and global climate change pose a great challenge to our environmental security. As we move into a global world from an international world, where the world's affairs were affected primarily by interactions between States, we face great challenges to the human security and sustainable development.

Space science and technology can assist us in facing many of the challenges that lie ahead. The Third United Nations Conference on the Exploration and Peaceful Uses of Outer Space, known as UNISPACE III, stressed how useful space applications can be in ensuring human development in a sustainable manner.[1] Convened in July 1999 in Vienna under the theme "Space benefits for humanity in the twenty-first century", UNISPACE III aimed at promoting greater use of space science and technology, in particular by developing countries, in all areas of scientific, economic, social and cultural development. The Conference discussed a wide range of issues related to the promotion of sustainable development and enhancement of the human condition. Civil society as well as future leaders of the world contributed to the discussion. The outcomes of UNISPACE III reflect the commitment made by governments and the collective wisdom of civil society and future leaders. The recommendations of UNISPACE III include some of the inputs provided by representatives of non-governmental organizations and industry as well as young professionals and university students.

By adopting "The Space Millennium: Vienna Declaration on Space and Human Development", UNISPACE III made an important contribution to the efforts of the United Nations to face the challenge of globalization and work towards an equitable global society. The Vienna Declaration identifies priority areas where space science and technology can assist in solving regional and global problems. It also contains the common strategy to be pursued by the

---

[1] See A/CONF.184/6, *Report of the Third United Nations Conference on the Exploration and Peaceful Uses of Outer Space.*

global society in expanding the use of space science and technology through international cooperation.

The United Nations Committee on the Peaceful Uses of Outer Space (COPUOS) and the Office for Outer Space Affairs have already begun to take concrete steps to turn the expectations and possibilities expressed at UNISPACE III into reality. The plan of action prepared by the Office and endorsed by COPUOS this year outlines the activities that the Office would carry out to implement the recommendations of UNISPACE III.[2] The Inter-Agency Meeting on Outer Space Activities, which serves as a focal point for inter-agency cooperation in space-related activities within the United Nations system, also joined the efforts of COPUOS and the Office in promoting the use of space technologies by organizations of the United Nations for various development purposes.

UNISPACE III was a turning point in strengthening the role of the United Nations to meet the needs of the peoples of the world through the use of space science and technology. In his report to the Millennium Summit in September this year,[3] the Secretary-General stated that success in meeting the challenges of globalization ultimately comes down to meeting the needs of peoples. He identified the following as the challenges for the United Nations in the twenty-first century: ensuring freedom from want; ensuring freedom from fear; and leaving a healthy environment to future generations. Space science and technology can play an important role in undertaking some of the major initiatives proposed by the Secretary-General, such as to build bridges across the digital divide and to create a new ethic of global stewardship.

Perhaps UNISPACE III preceded the steps being taken by the United Nations in setting and acting on this vision for the next millennium. If all states and non-state actors around the world share the spirit of the Vienna Declaration and work together to take action, the vision for the Space Millennium as reflected in the Vienna Declaration can become a reality.

II. PROTECTING THE ENVIRONMENT

When the United Nations was established, its founders set out to promote freedom from fear and freedom from want. They could not have anticipated the urgent need that we face today to protect the Earth's capacity to sustain the lives of future generations. Environmental sustainability is everybody's challenge. Our goal must be to meet the economic needs of the present without compromising the ability of the planet to provide for the needs of future generations.

---

[2] The plan of action proposed by the Office for Outer Space Affairs aims to achieve the following objectives: i) strengthening the role of COPUOS and its subcommittees in promoting international cooperation in the use of outer space ; ii) initiating a capacity-building progamme in areas relating to space law; iii) strengthening the activities of the United Nations Programme on Space Applications; iv) promoting the use of space technologies within the United Nations system; v) strengthening the partnership with intergovernmental and non-governmental organizations and industry ; vi) initiating a public outreach programme and a programme for young people; and vii) strengthening publication and information services. It is now being considered by the United Nations General Assembly for its endorsement.

[3] A/54/2000, *We the peoples: the role of the United Nations in the twenty-first century: report by the Secretary-General.*

UNISPACE III focussed on scientific knowledge of Earth and its environment and practical applications of space science and technology in protecting the environment and managing natural resources. UNISPACE III recognized the increasing threats of rapid environmental changes, including climate change, deforestation, desertification and land degradation, further depletion of the ozone layer, acid rain and loss of biological diversity. Such changes would have a profound impact on all countries, yet many important scientific questions remain unanswered.

In his Millennium Report, the Secretary-General stressed that only governments can create and enforce environmental regulations but, as he indicated, it is impossible to devise effective environmental policy unless it is based on sound scientific information. Large gaps in our knowledge still remain, and there has never been a comprehensive global assessment of the world's major ecosystems. The Millennium Ecosystem Assessment proposed by the Secretary-General seeks to produce a truly comprehensive global evaluation of the condition of the five major ecosystems: forests, freshwater systems, grasslands, coastal areas and agroecosystems.

As UNISPACE III indicated, satellites can provide the synoptic, continuous and long-term global observation needed to understand the Earth's system more comprehensively to address issues such as global climate change and other anthropogenic impacts on the environment. UNISPACE III noted various applications of space technology in the areas of weather and climate forecasting and management of natural resources. UNISPACE III called for action, for example: to develop a comprehensive, worldwide, environmental monitoring strategy for long-term global observations building on existing space and ground capabilities; to improve the management of the Earth's natural resources by increasing and facilitating research and operational use of remote sensing data, enhancing the coordination of remote sensing systems and increasing access to and affordability of imagery; and to develop and implement the Integrated Global Observing Strategy to enhance access to and use of space-based and other Earth observation data. These actions, if undertaken, would contribute to the efforts of the United Nations in carrying out the Millennium Ecosystem Assessment.

The work of the Office for Outer Space Affairs would contribute to strengthening the efforts being made by the United Nations as a whole in this area. The plan of action of the Office, endorsed by COPUOS, includes the development of a programme to promote the use of Earth observation data by institutions in developing countries. Under this programme, the Office would carry out the following activities: i) surveys to identify ongoing national and regional development projects that could benefit from the use of optical, infrared or radar data and a needs assessment study to identify the type and coverage of satellite images required by the identified user institutions in developing countries; ii) establishment and distribution of a comprehensive list of distributors of data from Earth observation satellites, as well as of analysed information, including the models used; and iii) provision of satellite data and hardware/software to user institutions to initiate pilot projects or strengthen ongoing pilot projects to use Earth observation data. Those activities would be carried out together with interested space agencies and other space-related international and national institutions. In addition, within the framework of the United Nations Programme on Space Applications, the Office will contribute to the activities of the Committee on Earth Observation Satellites (CEOS), especially activities to promote contributions from developing countries to comprehensive long-term global observation data sets.

The work of the Inter-Agency Meeting on Outer Space Activities will improve cooperation among United Nations entities in the use of space science and technology for protecting the global environment and management of natural resources. At its annual session in February 2000, the Inter-Agency Meeting noted that UNISPACE III had stressed the importance of space technology for sustainable development, including protecting the Earth's environment and managing its resources. For that reason, the Inter-Agency Meeting expressed concern that the importance of space applications had not been sufficiently stressed in Agenda 21,[4] adopted at the United Nations Conference on Environment and Development (UNCED), held in Rio de Janeiro, Brazil, in 1992. The Inter-Agency Meeting, therefore, agreed that it should provide input into the work of the Commission on Sustainable Development towards promoting recognition of the value of space applications for sustainable development. It also agreed that it should contribute to any future review of relevant provisions in Agenda 21, including a review of chapter 40, "Information for decision-making", to be conducted by the Commission on Sustainable Development in 2001, and to an event that might be held in 2002, 10 years after UNCED.

III. USING SPACE APPLICATIONS FOR HUMAN SECURITY, DEVELOPMENT AND WELFARE

Various space applications can enhance human security, development and welfare and alleviate extreme poverty. Satellites are increasingly providing important information for early warning and mitigating the impacts of disasters, which can destroy the economic and social base for prosperity. Both remote sensing and communications satellites can help to save lives by playing a vital role in supporting or enabling disaster response activities, including gathering and dissemination of information during an emergency and the provision of back-up communications. Data from remote sensing satellites, in combination with other information, have been used successfully to monitor the environmental conditions for the emergence and outbreak of infectious diseases, thus reducing the risk of social and economic loss caused by loss of the labour force. In the area of communications and broadcasting, newly proposed or enhanced satellite services, including mobile telephony, data, imaging, video teleconferencing, digital audio, multimedia and global Internet access, can reduce the information gap that has widened between those who can use technologies to access more information more quickly and those who cannot.

In his Millennium Report, the Secretary-General noted that a wide digital divide still exists in the world and expressed his belief that this digital divide can and will be bridged. As a concrete demonstration of how bridges over digital divides can be built, he announced a new Health InterNetwork for developing countries. This network would establish and operate 10,000 on-line sites in hospitals, clinics and public health facilities throughout the developing world. It aims to provide access to relevant up-to-date health and medical information tailored to specific countries or groups of countries. Another initiative to build digital bridges is a United Nations Information Technology Service, called UNITeS, a consortium of high-tech volunteer corps. UNITeS would train groups in developing countries in the use of and opportunities from information technology and stimulate the creation of additional digital corps. The Secretary-General also announced a

---

[4] *Report of the United Nations Conference on Environment and Development, Rio de Janeiro, 3-14 June 1992* (United Nations publication, Sales No.E.93.I.8 and corrigenda), vol. I: *Resolutions Adopted by the Conference, resolution I*, annex II.

programme which would demonstrate how the Information Revolution can benefit the United Nations itself. A new disaster response programme, "First on the Ground", would provide and maintain mobile and satellite telephones as well as microwave links for humanitarian relief workers.

The implementation of the recommendations of UNISPACE III would support and complement the efforts of the United Nations to enhance disaster response operations, to provide medical and health services and education opportunities to remote and rural areas, and to enable people around the world to benefit from the information revolution. UNISPACE III called for actions, for example: i) to implement an integrated global system to manage natural disaster mitigation, relief and prevention efforts through Earth observation, communications and other space-based services; ii) to improve public health services by expanding and coordinating space-based services for telemedicine and for controlling infectious diseases; iii) to improve knowledge sharing by giving more importance to the promotion of universal access to space-based communication services and by devising efficient policies, infrastructure, standards and applications development projects; and iv) to improve the efficiency and security of transport, search and rescue, geodesy and other activities by promoting enhancement of, universal access to and compatibility of space-based navigation and positioning systems.

In the area of disaster management, COPUOS and the Office for Outer Space Affairs took concrete steps towards implementing the recommendations of UNISPACE III. Following the agreement of COPUOS reached in June 2000, its Scientific and Technical Subcommittee will begin its consideration of an agenda item entitled "Implementation of an integrated, space-based global natural disaster management system" in 2001. This item would be considered in accordance with the following three-year work plan:

2001	Review of the types of natural disasters being faced and the extent of the application of space-based services being utilized for their mitigation;
2002	Review of existing and proposed satellite and data distribution systems that can be used operationally for disaster management and identification of gaps in those systems;
2003	Review of possible global operational structures to handle natural disaster management, making maximum use of existing and planned space systems.

For the work to be conducted by the Scientific and Technical Subcommittee in 2001, COPUOS agreed that CEOS should be invited to make a presentation and requested the Office for Outer Space Affairs to prepare a comprehensive preparatory document. COPUOS also agreed that domestic satellite operators of Member States and intergovernmental organizations involved in satellite communications should be invited to participate in the work of the Subcommittee on this issue.

Disaster management is one of the priority themes of the United Nations Programme on Space Applications. In order to support the work of COPUOS and the Scientific and Technical Subcommittee, the Office for Outer Space Affairs proposed, in its plan of action, to initiate a programme to promote the use of satellite communications and Earth observation data for disaster management by user institutions in developing countries. Under this programme, the Office would carry out the following activities: i) development of a series of modules for integrating space technology in disaster-specific management, consisting of technical, administrative and policy requirements and a step-by-step procedure to incorporate space technology into disaster management in a functional manner, including training required for officials and staff of civil protection authorities; and ii) development of plans and proposals to implement pilot projects in developing countries to use space technology in disaster management.

In addition, the Office for Outer Space Affairs increased its efforts to strengthen partnership with other United Nations bodies in areas relating to disaster management. The Office and the Secretariat of the International Strategy for Disaster Reduction (ISDR) will support each other's activities, such as the workshop on disaster management that the Office will organize in Chile in November 2000 and the workshop that the Office and the ISDR secretariat will organize in Africa in 2001. In the area of humanitarian response to natural disasters, the Office contributed to the Secretary-General's report on the strengthening of the coordination of emergency humanitarian assistance of the United Nations and stressed the usefulness of space applications in disaster response operations. Upon invitation by the Office for the Coordination of Humanitarian Affairs, the Office for Outer Space Affairs participated in the Panel on Natural Disasters during the Humanitarian Segment of the Economic and Social Council, which took place in New York from 19 to 21 July 2000.

The Office for Outer Space Affairs also plans to launch training modules, in particular for developing countries, in other areas of space applications that would improve human security, development and welfare. Those training modules would consist of workshops and seminars to be followed by support for pilot and demonstration projects conducted by the successful participants of the workshops and seminars. One is a training module to make effective use of satellite communications for distance education, to promote literacy and to reduce information gaps. Another is a module to build capacity, particularly of developing countries, to use global positioning and navigation satellite systems in their research and operational activities; the Office also proposed to launch a training module consisting of workshops or seminars that may be followed by pilot projects. As part of its strategy to re-orient the long-term fellowship programme and training courses to further promote sustainable development, the Office also plans to launch a programme called "Technology Outreach Programme". This new programme aims to promote the successful use of knowledge acquired by university educators who have participated in the training courses organized by the United Nations Programme on Space Applications. By providing grants, the programme would support ongoing pilot projects and practical demonstrations conducted by those university educators in their education curricula.

IV. ENHANCING EDUCATION AND TRAINING OPPORTUNITIES FOR PRESENT AND FUTURE GENERATIONS

Whether space science and technology can better the human condition, particularly in developing countries, depends most importantly on the availability of human resources with sound knowledge and skills. The whole process of capacity-building involves not only research, education and training but also establishing policies, establishing institutional frameworks and physical infrastructure, ensuring funding support and gaining experience. This is the reason why the Office for Outer Space Affairs, through its United Nations Programme on Space Applications, has continued to provide education and training opportunities for researchers and decision makers in developing countries in various areas of space applications and technical advisory services to assist developing countries in creating an environment conducive for space-related projects.

One of the major activities of the United Nations Programme on Space Applications has been to provide support to the regional centres for space science and technology education, affiliated with the United Nations, including the Network of Space Science and Technology Education and Research Institutions for Central Eastern and South-Eastern Europe. UNISPACE III identified this activity as a key element of the effort to build capacity in developing countries to develop and use space science and technology. As indicated in its plan of action, the Office for Outer Space Affairs intends to strengthen its support to the regional centres and the Network, in particular their operational activities, by seeking financial and other support from governments, international development financial institutions, space agencies, universities and specialized space-related institutions as well as industry.

Our efforts for capacity-building in space science and technology would not be complete without providing education and training opportunities for the young. More than 1 billion people today are between the ages of 15 and 24, and nearly 40 per cent of the world's population is younger than the age of 20. Young people are a source of creativity, energy and initiative. Given the chance to learn and work, they can contribute significantly to social and economic development. As future leaders of the world, they hold the key for the prosperous future of human civilization. In his Millennium Report, the Secretary-General also emphasized the importance of education. He invited governments, for example, to meet the objective that by 2015 all children should complete a full course of primary education.

UNISPACE III recognized the important role that young people can play in the coming years and stressed the need to strengthen efforts to provide young people with opportunities to express their unique and innovative ideas and visions for space activities. The Space Generation Forum was, therefore, organized during UNISPACE III by and for young professionals and university students to express their visions and perspectives of future space endeavours.

The Space Generation Forum was a major success of UNISPACE III. Some of its recommendations were included in the recommendations of UNISPACE III, such as the creation of a consultative mechanism, within COPUOS, to facilitate the participation of young people in space-related activities. COPUOS this year agreed that the Youth Advisory Council, which had been established as a voluntary body succeeding the Space Generation Forum, could be granted observer status with COPUOS. In addition, following an agreement by COPUOS, the Scientific and Technical Subcommittee will consider at its session in 2001 government and private activities to promote education in space science and engineering.

The Office for Outer Space Affairs also plans to provide more opportunities for young people to learn about space activities by organizing workshops and seminars, within the framework of the United Nations Programme on Space Applications, for young professionals and university students on various scientific and technological, legal and policy issues relating to space activities. In addition, the Office plans to launch a programme of visits by astronauts, cosmonauts and space scientists and a programme to provide opportunities for students of primary and secondary schools, in particular those in developing countries, to participate in ongoing human space flight activities. Through its International Space Information Service, the Office also plans to develop and carry out an interactive multimedia education service in particular for students of primary and secondary schools. This multimedia education service would include dissemination of CD-ROMs containing multimedia educational materials on various space-related subjects, creation of a public on-line question-and-answer forum and organization of on-line interviews with astronauts and cosmonauts.

V. CONCLUSION

It is a goal of the United Nations to ensure that all people in the world can lead a dignified life. It is in the name of all people that the Charter was written. Realizing their aspirations remains the vision of the United Nations.

UNISPACE III showed how space science and technology can help meet the needs of people around the world. It highlighted the problems which need urgent attention and action by the international community. We need to find out more about Planet Earth so that we have a clear idea about how our activities affect the global climate and the environment, and we can agree on which of our activities need adjustment or correction. We need to ensure that advances in space applications help the poorest and that the most vulnerable people of the world have access to the basic necessities of life. We need to ensure that space technology helps people around the world to benefit from globalization.

The expectations and possibilities expressed at UNISPACE III would not become reality unless concrete actions are taken by all concerned. With the Vienna Declaration and the specific actions identified at UNISPACE III, the question is not what action to take, but when to take action.

The Committee on the Peaceful Uses of Outer Space and the Office for Outer Space Affairs have already begun to take action to implement recommendations of UNISPACE III to yield tangible results in the near future. Some non-governmental entities are also joining their efforts. If many other States and non-governmental entities support the work of the Committee in promoting the use of space tools for improving the living condition of people of the world, in particular those in developing countries, we can translate our vision for the Space Millennium into reality, and we can create the equitable global village that our future generations deserve.

SPACE ASTRONOMY, CURRENT MISSIONS AND TRENDS FOR THE NEXT MILLENNIUM[*]

Marie-Jose Deutsch
Jet Propulsion Laboratory, California Institute of Technology, Pasadena, CA, U.S.A.

ABSTRACT

Space astronomy allows access to wavelength regions that are not available to ground-based observatories. Collecting and analyzing radiation emitted by phenomena throughout the entire electromagnetic spectrum, the "Four Great Observatories" are performing astronomical studies over many different wavelengths and overlapping in time enabling concurrent observations. The Chandra X-Ray Observatory, deployed in July 1999, will observe x-ray images and spectra of violent, high temperature events and objects to help us understand black holes, quasars, and high temperature gases. The Space Infrared Telescope Facility (SIRTF) will launch in December 2001. It will be capable of observing in the near infrared in the 3 to 180 micron range and provide for imaging, photometry as well as spectroscopy. The primary science themes are the detection and study of brown dwarfs and super-planets, protoplanetary and planetary debris disks, ultraluminous galaxies and active galactic nuclei, and deep surveys of the early universe. The detector arrays offer orders of magnitude improvements in capability over past infrared detectors.

Astronomical missions scheduled for 2005 and beyond are enabled through advanced technology development. The Space Interferometry Mission (SIM) will use optical interferometry technology, while The Next Generation Space Telescope (NGST) will require large, ultra-light, and deformable mirrors, and very sensitive instruments. SIM will determine the positions and distances of stars several hundred times more accurately then any previous program. This will allow SIM to probe nearby stars for Earth-sized planets. SIM will also pioneer a technique to block out the light of bright stars to take images of areas close in to the stars. NGST is to be launched in 2007 and will study how galaxies evolve, how stars and planetary systems form and evolve, and what the life cycle of matter is in the Universe.

SIRTF, SIM and NGST are part of NASA's Origins program and Chandra is part of NASA's Structure and Evolution of the Universe program. SIRTF and SIM are managed for NASA by the Jet Propulsion Laboratory (JPL), California Institute of Technology. NGST is managed by NASA's Goddard Space Flight Center and Chandra is managed by NASA's Marshall Space Flight Center.

---

[*] *This paper was presented at the "Ninth United Nations/European Space Agency Workshop on Basic Space Science", held from 27 to 30 June 2000 in Toulouse, France, and does not necessarily reflect the views of the United Nations.*

INTRODUCTION

This paper expands on the current and near future NASA Space Astronomy Missions as well as the plans for beyond 2005.

Space astronomy allows access to wavelength regions that are not available to ground-based observatories due to spectral absorption by the Earth's atmosphere. Continuous dark times give space observatories long continuous exposure times. In the infrared, space based observatories provide an additional bonus in that they are free of black body radiation of the Earth's atmosphere and telescopes can be cooled down to very low temperatures, which increases the sensitivity a thousand fold, giving a million-fold increase in the speed of the observations. Wavelengths from gamma rays to far-infrared are available, sometimes on the same spacecraft.

GREAT OBSERVATORIES

The "Great Observatories", a NASA program, is a set of four large space-born observatories, designed to study the universe over many different wavelengths. The four missions have overlapping operations phases to allow for concurrent and follow-up observations of features at many different wavelengths. The four missions that are part of the Great Observatories program are the Hubble Space Telescope (HST), the Compton Gamma-Ray Observatory (CGRO), Chandra X-Ray Observatory (CXO), and the Space InfraRed Telescope Facility (SIRTF). HST was launched in 1990 by the NASA Space Shuttle. Subsequently it has been refurbished with a number of new instruments and capabilities. More servicing missions to HST are planned. HST observes the Universe at ultraviolet, visual, and near-infrared wavelengths. HST has contributed to understanding of the expanding universe and to determining the age of the universe to be 12 billion years. CGRO was deployed by the Space Shuttle in 1991. Its mission is to collect data on some of the most violent physical processes in the universe, characterized by their extremely energetic emissions. CXO was deployed in 1999 and is observing such phenomena as black holes, quasars, and high temperature gases in the x-ray portion of the electromagnetic spectrum. And SIRTF, to be launched in 2001, will conclude the series of Great Observatories by providing coverage in the infrared part of the spectrum.

CHANDRA X-RAY OBSERVATORY

The Chandra X-Ray Observatory studies the violent universe, phenomena that release their enormous amounts of energy at x-ray wavelengths, temperatures that reach millions of degrees, high velocity gases accelerated by gravity to nearly the speed of light, and magnetic fields that are a trillion times stronger than the Earth's. Chandra is designed to study the deaths of stars, collision of galaxies, explosion of stars that release heavy elements from their interiors, black holes and quasars or neutron stars.

Imaging of star formations and young stars with Chandra helps scientists understand the evolution of life on Earth. When massive stars or supernovae erupt, they release their content of heavy elements and energy into the surrounding space. Shock waves, formed during this explosion due to the collision of ejecta from the supernova with the circumstellar material, heat it up to 10 million degrees. Like sonic booms, secondary waves resulting from this collision further stir up the surrounding gas. The bright remnant in the center of the explosion may be a neutron star or black hole. Thus heavy elements such as carbon, nitrogen, oxygen and iron, necessary for life, formed inside the massive stars, help seed the interstellar gas, heat it with their energy of their radiation, stir it up with their force of the blast waves and cause new stars to form.

Other x-ray sources are caused by super-hot gas and dust particles that are being swallowed up by black holes. Accretion rate by quasars, due to the tremendous gravity pull, speeds up particles to form a rotating disk. Collision between particles heats them to many millions of degrees producing x-rays. The study of the x-ray energies allows the determination of the particle motion as they approach the black holes and gives scientists information about the gravity fields around black holes.

Most galaxies in our universe are grouped together in clusters. These clusters are filled with large clouds of multimillion-degree gas, held together by gravity and visible in the x-ray spectrum. However, the mass of the galaxies and gas together is not sufficient to provide the gravity to hold the clusters together. So the theory emerged that an additional mass of dark matter must exist to explain the phenomena. This in fact would one lead to conclude that most of the mass of the universe may be dark matter collapsed into stars, planets and black holes that are not observable to us directly. The study of the hot gas by Chandra might shed some light on the dark matter puzzle.

Chandra was launched July 23, 1999, from Shuttle Columbia into a highly eccentric orbit with a maximum altitude of 140,000 km. The orbit is 64 hours long with 14 hours in the high radiation environment of the Earth, when the telescope instruments have to be turned off to avoid damage. The telescope is being operated from Harvard-Smithsonian Observatory, the first mission to be operated at a site other than a NASA center.

The telescope system has very specialized technology. Cylindrical mirrors have replaced normal mirrors, which would be penetrated by the x-rays. Two sets of four nested mirrors allow the incoming x-rays to ricochet off the paraboloid surface and hyperboloid surface (Figure 1) to be funneled toward the center detectors. The mirrors are made of Zeodur material with a 600 Angstrom of iridium coating to create the smoothest surfaces ever, thus allowing the x-rays to graze off its highly polished mirror surfaces. The entire telescope is wrapped in a multi-layer insulation controlling the temperature inside the telescope, preventing expansions and contractions of the mirrors and ensuring greater accuracy of the observations.

Chandra's instruments can serve both as imager or spectrometer. The High-Resolution Camera (HRC) records the x-ray images between 0.1 and 1 keV on its microchannel plate detectors. The camera is composed of two detectors, one optimized for imaging and the other optimized for spectra. The imaging field of view is 30x30 arcmin with a spatial resolution of approximately 0.5 arcsec. The HRC complements the second instrument, the Charged Coupled Imaging Spectrometer (ACIS). ACIS' two CCD arrays can provide simultaneous time-resolved imaging and spectroscopy of up to 50 different energies within the range of 0.2 and 10 keV. The imaging field of view is 16x16 arcmin. In order to gain even more energy information, two screen-like instruments, called diffraction gratings, are inserted between the telescope and the detectors. The gratings change the path of the x-rays, depending on their color or energy, and the x-ray camera records the color and the position. The High Energy Transmission Grating (HETG) is used on high to medium energies ranges of 0.4 to 10 keV with a spectral resolution, E/delta (E), ranging from 60 to 1000. The Low Energy Transmission Grating (LETG) disperses low energies from 0.08 to 2 keV, with a resolution of 30 to 2000. The study of such a large variety of energies in the x-ray spectrum enables scientists to determine the composition of the objects emitting the radiation and learn about their evolution.

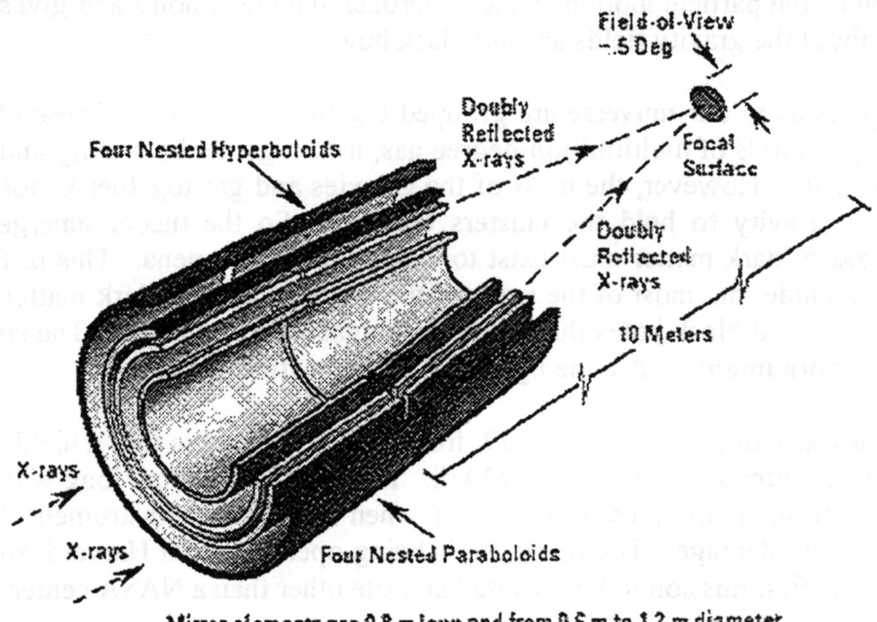

Figure 1. Chandra's special mirror configuration.

STUDYING THE UNIVERSE IN THE INFRARED – SPACE INFRARED TELESCOPE FACILITY

Most objects in the universe, such as cool stars, planets, interplanetary dust, interstellar gas and the universe itself are too cool to emit visible light, but they do emit in the infrared. Interstellar dust that shrouds the center of galaxies and clusters is totally opaque in the visible and ultraviolet, but becomes mostly transparent in the infrared, allowing scientists to study regions of star formation. The expanding universe has objects in the distant universe that move away from us. This cosmic expansion shifts the light from distant galaxies into the infrared. SIRTF, a cooled meter-class telescope, with background limited detectors and multiple instruments, is going to offer substantial improvement over past and existing capabilities.

Some of the scientific themes to be pursued by SIRTF are identified below and are expected to greatly contribute to our understanding of these phenomena.

1) Objects such as brown dwarfs and super planets have too little mass to ignite the fusion reactions of true stars, but they are larger and warmer than planets found in our solar system. SIRTF will allow detection and characterization of the abundance of these objects which can be found in the halo of galaxies and in young star clusters, contributing to the understanding of the dark matter that is thought to exist in the universe. In addition spectroscopy will contribute to the determination of their nature and composition.

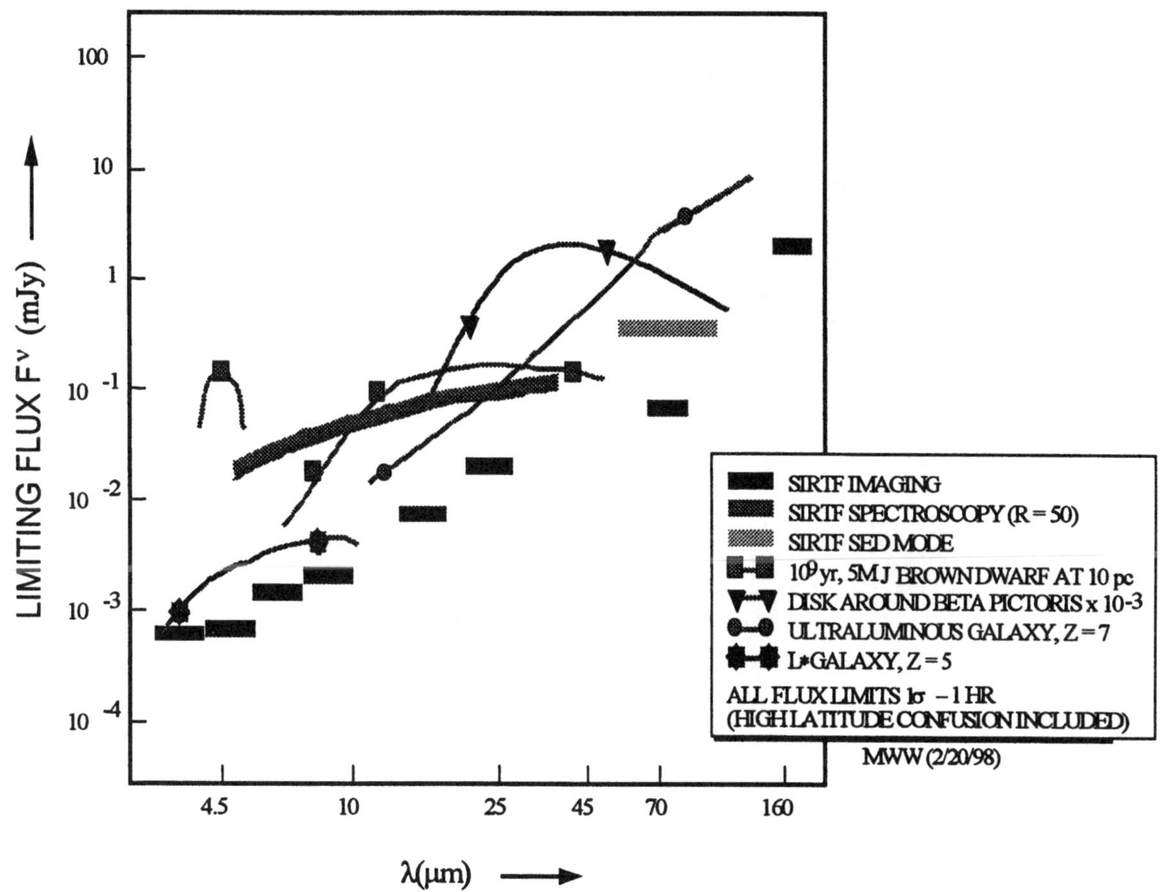

Figure 2. SIRTF Sensitivity

2) The properties and interrelationships of comets, asteroids, interplanetary dust and other small bodies in the Solar System are to be investigated by SIRTF. The structure and composition of disks of debris and dust surrounding nearby stars may be an indicator of solar system formation. SIRTF will be able to trace the evolution of these formless clouds of dust and gas into a mature solar system like ours. By comparing low resolution spectra of the disk material with emission spectra from comets, SIRTF provides comparisons of the primitive solar system, of which comets are thought to be remnants, with those of the debris disks around other solar systems.

3) The study of the physical processes which power the Ultraluminous Galaxies, which are at the core of intense bursts of star formation, stimulated by colliding galaxies and the probing of dust enshrouded quasars and active galactic nuclei powered by black holes, will be at the center of SIRTF's inquiry. In addition, spectroscopy will be able to establish the source and abundance of heavy metals detected in the red shift.

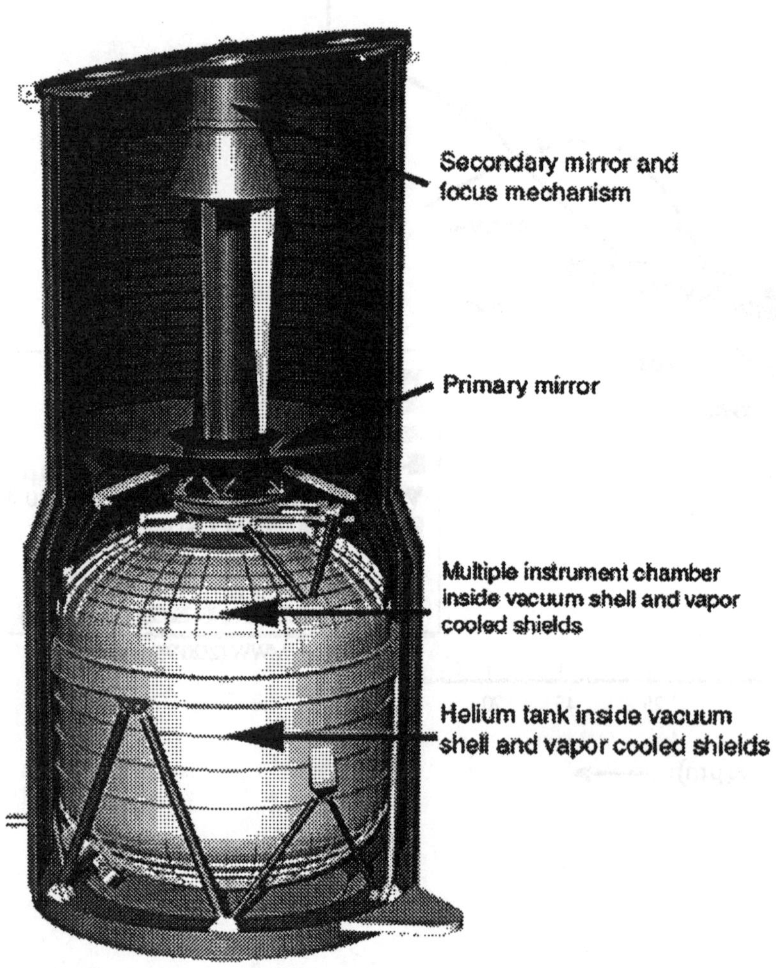

Figure 3.   SIRTF Warm Launch Architecture.

4) Deep surveys of the distant and early Universe will be explored at red shifts greater than 3. SIRTF is expected to discover thousands of galaxies per square degree on the sky.

5) In addition to these themes, SIRTF is also part of the *Origins* Program, which seeks to understand the origins of the Universe, galaxies, stars and planets. Thus SIRTF will address a wide range of astronomical investigations, including studies of the outer Solar System, the early stages of star formation, and the origin of chemical elements.

SIRTF will be launched in December 2001 into an Earth trailing heliocentric orbit, drifting away from Earth at a rate that brings it out to .64 AU in 5 years. This orbit is favorable for both mass and thermal reasons and has considerably helped simplify and reduce the development cost of SIRTF. The solar orbit provides such important advantages as elimination of occultations and eclipses providing continuous sky access and excellent visibility.

SIRTF has taken advantage of considerable infrared detector technology improvements brought about by industry, which served military interests in developing detectors for high-background temperature environments in wavelengths shorter than 30μm. Astronomers have adapted this technology for low-background and high sensitivity sky observations in the infrared up to 200 μm. Thus SIRTF will feature a thousand-fold increase in sensitivity. The size of the arrays has increased many thousands of times as well. The benefit of this detector revolution is that observations can now be accomplished in very short integration times considerably increasing the observatory's efficiency. SIRTF features three instruments, which provide imaging, photometry and spectroscopy over wavelengths ranging from 3.6 μm to 160 μm. Figure 2 shows the limiting flux as a function of wavelength, an indication of how faint an object SIRTF is able to observe with its complement of detectors.

SIRTF will be launched at ambient temperature and allowed to cool radiatively in space. Only the instrument detectors and the compact cryostat are encased in a vacuum shell sitting on top of 360 liters of superfluid helium tank that cools the instrument chamber (Figure 3). Within one week of passive cooling the temperature of the observatory outer shell will fall to about 50 °K. The helium boil-off will further cool the inner telescope assembly, including the instruments, down to 5.5 °K. The liquid Helium bath will serve as a heat sink and will remain at 1.5 °K. The outer shell temperature will reach about 33 °K.

NASA'S ORIGINS PROGRAM

The *Origins* program is a NASA program that looks into the next millennium and defines the high-level exploration goals between 2000 and 2015. *Origins* program seeks to understand the origins of the universe, galaxies, stars and planets. How did galaxies form in the early universe and what role do galaxies play in the appearance of planetary systems and life? How do stars and planets systems form and are there life-sustaining planets around other stars? How did life originate on Earth and does it exist elsewhere in the Universe?

These three leading questions have led to the definition of a number of mission objectives, which drive the astronomy missions for the new millennium.

1) The new millennium missions seek to understand the role gravity plays in the emergence of galaxies from the smooth particle distribution of the early Universe. These missions study the birth and aging of a galaxy and how this process influences the chemical composition of stars, planets, and living organisms. The Next Generation Space Telescope (NGST) mission will be designed to achieve this goal.

2) A number of missions such as NGST, Space Interferometry Mission (SIM) and Terrestrial Planet Finder (TPF) will be looking for planetary systems around young stars and for life sustaining planets around other stars close to our solar system.

3) Rovers, aerobots and in-situ instruments will be designed to study how life arose elsewhere in the solar system and beyond. Bioastronomy will investigate if life exists elsewhere in the Universe and draw parallels from life on Earth.

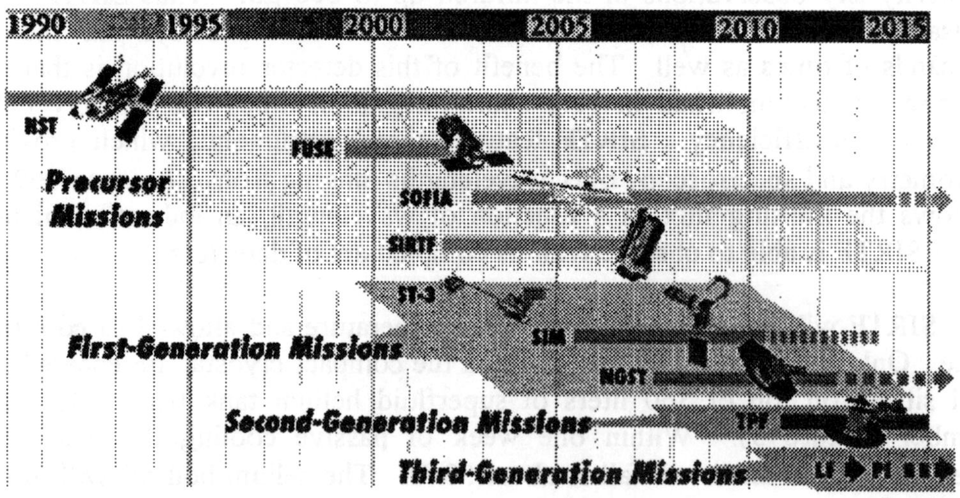

Figure 4. *Origins* Timeline

The technological feat to accomplish the *Origins* science goals is enormous. The timeline (Figure 4) shows how chronologically the missions are grouped in order of increasing technological challenge. The **Precursor** missions, which SIRTF is part off, are technologically very challenging but they are well along in their development and/or have become operational. The **First Generation** missions require revolutionary new technologies that are currently under development. These include much larger but lighter optics and collections of telescopes that together make images sharper than very large telescopes. These missions will be launched after 2005. SIM and NGST are part of this group of missions. They are part of technology pathfinders to the **Second Generation** missions. TPF is currently the only mission planned in this latter category and will be launched after 2010. And finally the **Third Generation** missions will have to deal with such an enormous technological challenge that at this point these missions are only a vision. Planet Imager (PI) is the only mission planned in this category in the time frame after 2015.

THE SPACE INFEROMETRY MISSION (SIM)

SIM is a space based optical interferometer for precision astrometry. The design is a 10-meter baseline, Michelson beam combiner to be launched mid-2005, for a minimum 5-year mission lifetime. SIM will be performing global astrometry, local astrometry, synthesis imaging and will demonstrate the new technology of fringe nulling for future missions. SIM will be observing in the band-width between 400 and 1000 nm. The expected all-sky astrometry accuracy to be achieved by end of mission is 4 μas. SIM will measure the proper motion by parallax with an accuracy of 2 μas/yr and the narrow-angle astrometry accuracy is expected to be 1 μas in one hour for objects at a distance of 10 pc.

The primary science objectives of SIM are to search for planets around stars within 150 pc, measure distances to stars throughout the Galaxy and demonstrate fringe nulling technology for future interferometry missions. The following science investigations are enabled by this mission:
1)   Search for astrometric signature of planets around nearby stars.
2)   Measure distances to spiral galaxies using rotational parallaxes
3)   Study the mass distribution in the halo of our Galaxy
4)   Observation of the dynamics of our local group of galaxies
5)   Determination of the spiral structure of our Galaxy
6)   Calibration of the cosmic distance "latter"
7)   Determination of the ages of globular clusters
8)   Observation of the internal dynamics of globular clusters
9)   Measurements of the masses and distances to MACHOs
10) Determination of accurate masses for low-mass stars in binaries
11) Imaging of emission-line gas around black holes in active galactic nuclei
12) Imaging of dust disks around nearby stars through nulling

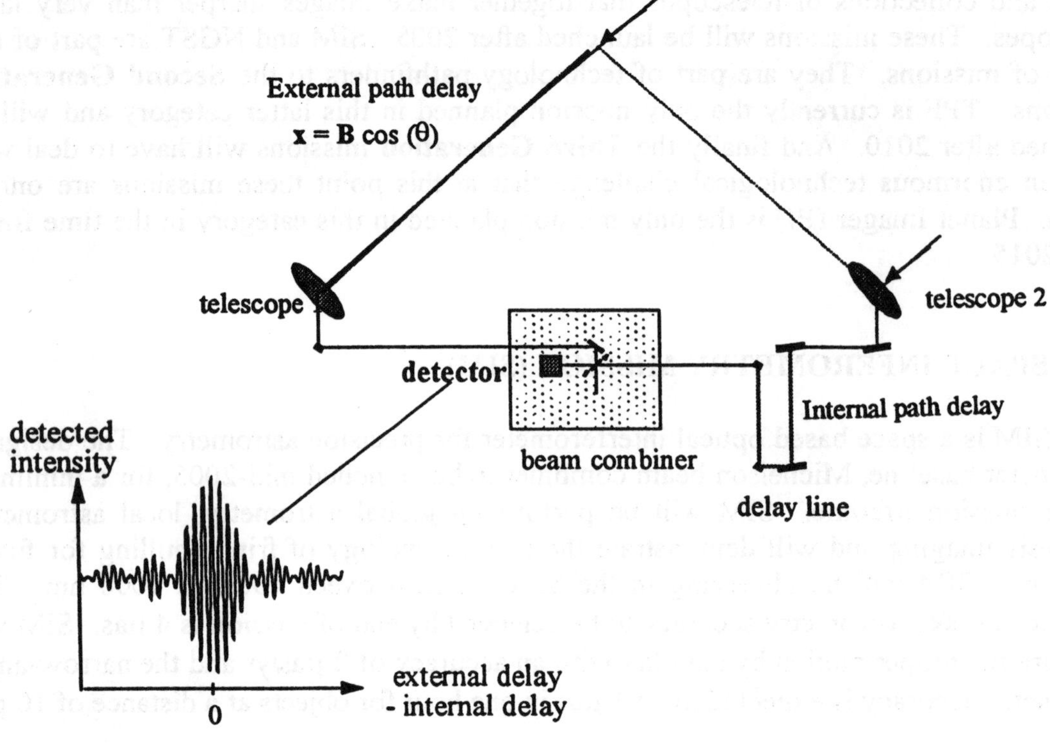

Figure 5.   SIM Astrometric Measurement – White light fringe position measures one coordinate of star position.

Figure 5 illustrates the SIM interferometer. It is composed of two collectors located at each end, an internal delay line and a beam combiner to interfere the starlight arriving from each interferometer arm. As the starlight enters the collectors it is sent toward the beam combiner in order to form white-light fringes, which occurs when the optical path traversed through the left arm is equal to the right arm. An optical delay line is added to one arm in order to shorten or extend the optical path as needed, until they are equal. Knowing the exact delay distance, and the exact distance of the baseline, the determination of the angle to the star is made.

However, in order to measure distances between stars, the space-based platform needs to remain stable at the µas level. This presents a major challenge when observing dim targets, which require very long integration times. Two additional interferometers (Figure 6) – called guide interferometers – are added to determine the orientation of the baseline vector and observe a pair of bright stars while the science interferometer observes a number of science targets. The spacecraft attitude control system senses from the guide interferometers the changing baseline orientation and corrects for its movement during the science observation period. White fringe acquisition and measurements will be limited by the changes in attitude of the spacecraft but also by the on-board disturbances which vibrate the elements in the optical system. An internal metrology system monitors the optical path vibrations. The laser metrology will feed data to the spacecraft, which will autonomously command the delay line to a position. Together the external baseline metrology and the internal metrology allow sufficient stabilization for long science target integration times.

Measuring the angle to a distant star from two vantage points (figure 7) allows the determination of the parallax angle π and the distance ($D = 1/\pi$ parsec) to that star. The bigger the baseline, the better the accuracy of the measurement. Knowing the distance to stars in turn provides input to determine the brightness of the star. This in turn leads to determination of the chemical composition and evolution of the star. Using parallax, SIM will measure distances to 25 kpc with an accuracy of 10%. SIM will also determine distances to star clusters up to 130 pc, the galactic center at 8.5 kpc and the large Magellanic cloud at 50 kpc.

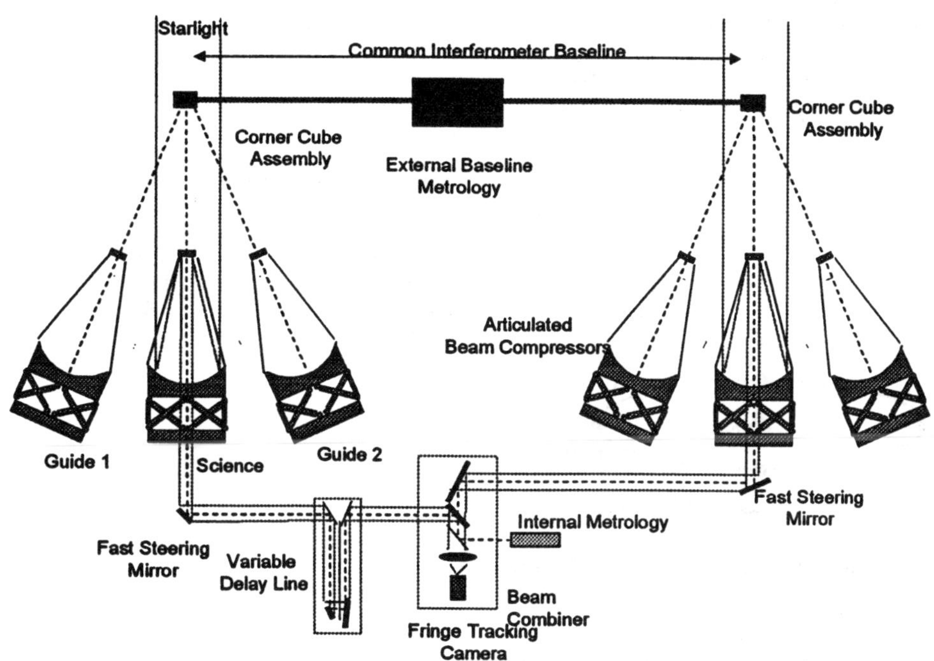

Figure 6. SIM Optical Layout – Triple Inferometer

A grid catalog, made up of quasars and other bright distant stars, will be established. Science observations are accomplished using these reference grid stars. Distances to spiral galaxies are performed using rotational parallaxes and ground based radial velocity measurements, allowing the astrometric measurement of galactic rotation. 3-D motions of a large sample or cluster of stars can thus be studied, their mass distribution determined, and information about formation and evolution deduced. Using stellar velocities, the size, mass distribution and dynamics of our galaxy is studied. The method to search for planets around other stars uses the astrometric detection of wobble due to gravitational tug of unseen planets. Earth-mass planets signature is 1650 times smaller than Jupiter size planets and is detectable only around the nearest stars at 10pc away.

As a conclusion, SIM will serve as a technology precursor for future interferometers in space. In addition to unprecedented science return, it will demonstrate the operation of a Michelson interferometer in space, fringe nulling technology, control of thermal and vibration environment, synthesis imaging in space, precision deployment and angle and pathlength control.

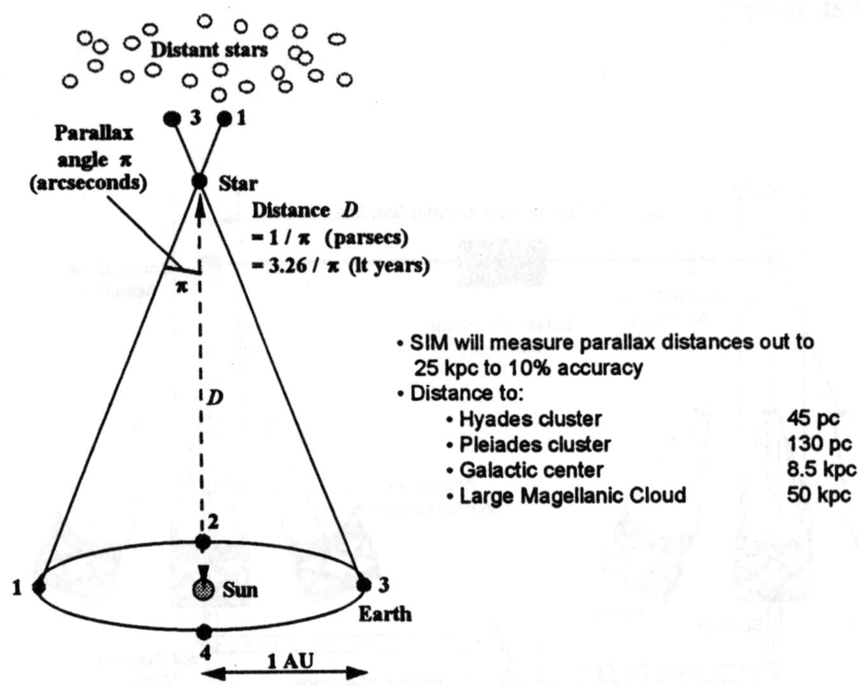

Figure 7.   Measuring Distances in the Galaxy

NEXT GENERATION SPACE TELESCOPE (NGST)

HST and confirmation from CGRO have established the age of the universe at 12 billion years. Telescopes have been peering farther and farther back in time and space to study objects from the earliest universe. HST discovered that galaxies had already formed 5 billion years after the big bang. The Cosmic Background Explorer (COBE) looking back as far as 1 to 100 million years from the big bang, shows only very slight differences in the density of matter, differences which are believed to have led to the formation of larger structures such as galaxies. NGST will be designed to investigate the time 1 billion years from the big bang, a time when primordial matter began to form into galaxies and stars.

NGST's science objectives are to study the birth of the first galaxies, the formation of stars and planets, the shape and evolution of the universe, the chemical evolution of the universe and the nature of dark matter. The target objects for these studies are deep fields, the universe at redshifts beyond $z > 2$, supernovae, stellar populations, cosmic distances and objects from the Kuiper Belt. Flux densities of these observations are in the range of nanoJanskys - Figure 8. These major science drivers have led to technology requirements for very large apertures, (ultralight 8 m deployable telescope) infrared capabilities, (cryogenic optics and radiatively cooled telescope) and very large detector arrays observing in the range from 0.5 to 30 μm with a cosmic background limited sensitivity. Due to limited integration and test on the ground of these very large structures, the mission and flight design must lend themselves to ease and flexibility in operations. The technology breakthroughs required for NGST have put its launch after 2008. The mission lifetime goal is up to 10 years.

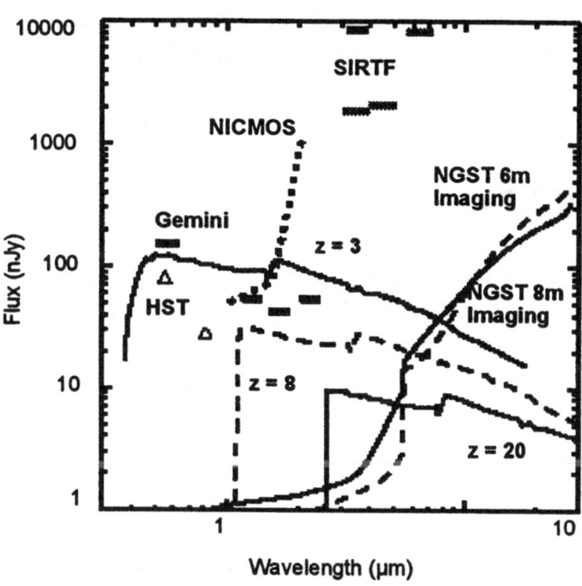

Figure 8.   Projected NGST Sensitivities - 10000 sec. Exposure, S/N = 10σ; λ/Δλ = 3

As is the case for all advanced NASA missions, the development strategy for NGST is to establish the main science design drivers and determine the technology requirements to implement those drivers. Based on available industry technology, heavily leveraging of military spin-offs and with new NASA funded stretch technologies, a detailed technical, cost and science trade is performed. From this trade study emerges a cost driven implementation plan and technology roadmap. More and more frequently international or other agency contributions are included in the mission trade space to make the mission feasible.

The technology roadmap becomes the critical path for the mission. Rapid prototyping shows the feasibility of key technologies required for the mission. A robust pre-development technology program with well-defined technology milestones is the next essential step. Ground testing and flight technology demonstrations validate the new technology. And finally the technology, if successfully validated in flight, is used to develop the ambitious science mission. Consequently the lead-time for this mission is very long and NGST won't be launched until late in the first decade of the new millennium.

CONCLUSIONS

The *Great Observatories* are leading the astronomical discoveries into the new millennium. They serve as scientific and technologic precursors to the much more ambitious missions under development for the *Origins* program, whose quest is to study the origins of the universe, galaxies, stars, planets and life itself. Challenges for those missions are many orders of magnitude larger and they are crucially dependent on the successful technology roadmap paving their future. Keeping the precarious balance between technological advances and mission simplicity is the great challenge coming with a bold science program.

ACKNOWLEDGEMENTS

This work was supported by the Jet Propulsion Laboratory, California Institute of Technology, under contract with the National Aeronautics and Space Administration. The author would like to acknowledge Joy Nichols for the information on Chandra provided toward this paper.

REFERENCES

Deutsch, M.-J., Bicay, M.D., eds. *Observing with SIRTF: Opportunities for the Scientific Community*. To be published in the proceedings of 8th UN/ESA workshop on Basic Space Science, Astrophysics and Space Science, Kluwer Academic Publishers, 1999.

Unwin, S., and Stachnik, R., eds. *Working on the Fringe: An International Conference on Optical and IR Interferometery from Ground and Space*. To be published in Astronomical Society of the Pacific Conference Series, 2000.

NEAR-EARTH ASTEROID OBSERVATION AND ITS EDUCATIONAL APPLICATION[*]

Syuzo Isobe
National Astronomical Observatory, 2-21-1, Osawa, Mitaka, Tokyo, Japan
and
Japan Spaceguard Association, Gloria Hatsuho 609, 2-3-14, Kyonan, Musashino, Tokyo, Japan

ABSTRACT

It has been said that opportunities for obtaining experience with scientific research in the real world are decreasing for high school pupils in many countries, especially in Japan. Therefore, some educational programs using astronomical observation data have been developed by different groups as one method to overcome these difficulties.

The Japan Spaceguard Association is building two new telescopes to detect near-earth asteroids (NEAs) that have the possibility of becoming global hazards. The telescopes will be fully operational by the end of 2000. One has an aperture of 1 m and a field of 3 degrees, covered by 10 2 k x 4 k CCD chips.

Using tela-byte data per month produced by this telescope, we are preparing an educational program to allow both schoolchildren and members of the public to detect asteroids and perhaps discover new asteroids. This gives them not only the chance to conduct practical scientific research but may also rub off some of the passion of the real scientific world.

1. HAZARDS OF COLLISION OF NEAR-EARTH ASTEROIDS

Recent studies (cf Gehrels 1994) showed that there are many asteroids that have a chance of colliding with the Earth and resulting in a local or global catastrophe. One can see evidence of this in the geological record. There are many craters on the lunar surface and also on the terrestrial surface. In addition, astronomical observations have discovered many asteroids (about 1,000 by May 2000) with pericentric distances less that 1.3 AU (astronomical units) from the Earth, which are called near-Earth asteroids (NEAs).

Once the orbital elements of an NEA are determined, one can make a numerical integration to evaluate its future possibility of colliding with the Earth. Then, a plan will

---

[*] *This paper was presented at the "Ninth United Nations/European Space Agency Workshop on Basic Space Science", held from 27 to 30 June 2000 in Toulouse, France, and does not necessarily reflect the views of the United Nations.*

be developed to mitigate or avoid collision. However, until we detect all candidate NEAs, there remains a possibility of the Earth being hit by an undiscovered NEA.

An important first task is to detect all NEAs. There are NEAs with different sizes and it is too difficult to detect all the small asteroids (with diameters between 10-100 m). To detect 100-500 m asteroids, we need about 100 ground-based, space-based or lunar-based telescopes with apertures between 1 m and 10 m. This would be a very expensive undertaking, as is usual for scientific projects, and therefore further discussion is needed.

Collision of an NEA with a diameter larger than about 0.5 km (hereinafter referred to as a "large NEA") has a high possibility of resulting in either the extinction of humankind, just as dinosaurs became extinct 65 million years ago, or at least cultural destruction. We should certainly work to prevent at least this event. Fortunately, 1 m telescopes have the ability to detect all the large NEAs, but to do this within a reasonable period (about 20 years) we need 10-20 1 m telescopes dedicated to the NEA survey.

2. TELESCOPES FOR DETECTION OF NEAR-EARTH ASTEROIDS

There are four teams working actively on NEA detection in the United States of America (Table 1). They had discovered over 400 large NEAs by May 2000, a total which is expected to include nearly half of all existing large NEAs. Since NEA hazards are not only a United States problem but also an international problem, we need more telescopes at different longitudes and in both hemispheres. This is necessary to detect all large NEAs and to make follow-up observations, since an NEA usually moves quickly in the sky and may travel from the Northern Skies to the Southern Skies and vice versa. Therefore, we need more telescopes to detect and follow-up NEAs in different places.

Table 1. NEA detection teams in the United States

Program	Telescope	Site	CCD	Field of view (degree x degree)
Space watch	88 cm f/5 Newtonian	Tucson, Arizona	2 k x 2 k	0.6
	1.8 m f/2.7		2 k x 2 k	0.6
NEAT[1]	100 cm f/2.15 Cassegrain	Maui, Hawaii	4 k x 4 k	1.6
LONEOS[2]	59 cm f/1.91 Schmidt	Flagstaff, Arizona	4 k x 4 k	3.2
LINEAR[3]	100 cm f /2.1 Cassegrain	Socorro, New Mexico	2.5 k x 2 k	2

[1] Near Earth Asteroid Telescope
[2] Lowell Observatory Near Earth Observation System
[3] Lincoln Laboratory Near Earth Asteroid Research

In Japan, we have set-up a non-profit organization called the Japan Spaceguard Association and made a concerted effort to obtain financing for a new NEA observation center. Fortunately, we were able to obtain financial support from the Science and Technology Agency covering the fiscal years from 1998 to 2001, and are building two telescopes. A half-meter telescope, established at a new observatory called the Bisei Spaceguard Center in January, 2000, (figure 1) is making test observations and will be in full operation soon. A one-meter telescope is under construction, and will be on-site by autumn 2000 and in full operation by the end of 2000 (figure 2).

Figure 1. A view of the half-meter telescope at the Bisei Spaceguard Center.

Figure 2. An overview of the Bisei Spaceguard Center. The half-meter telescope is set up in a sliding roof and the one-meter telescope will be set up in a dome.

Since the Japan Spaceguard Association has written several papers on these telescopes (Isobe 1999, 2000a, 2000b), I will repeat this information here only to show the abilities of these telescopes relevant to educational applications. The half-meter and one-meter telescopes have very wide fields of 2 degrees and 3 degrees and liquid nitrogen cooled 2 and 10 times 2 k x 4 k CCD chips, respectively (Figure 3). The CCD detector systems have a quick read-out time (about 13 seconds) and can make an exposure time up to 40 minutes with small dark noise. Our estimations are that limiting magnitudes for a one-minute exposure gives 19 and 20.5 magnitudes for the half-meter and one-meter telescopes, respectively.

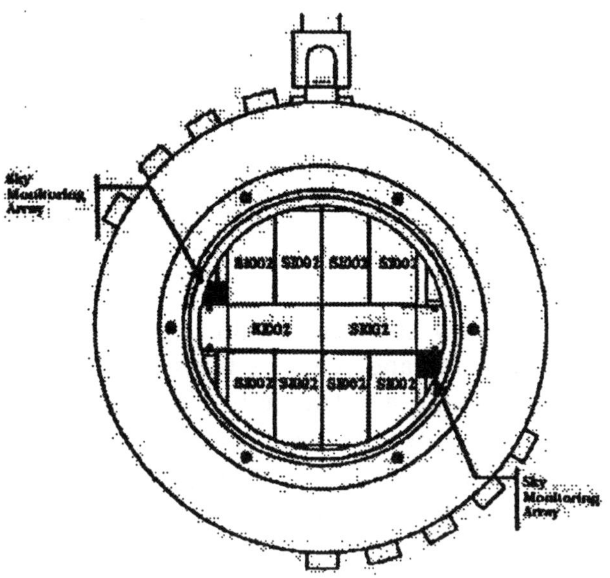

Figure 3. Distribution of a CCD array for a 3-degree field

We will expose each field of observation five times, covering about 700-1,000 square degrees per night. This will generate large amounts of data, by which we can detect about 10,000 asteroids, although the exact number depends on ecliptic latitude. This large number makes it possible for us to construct a new educational program.

3.   AN EDUCATIONAL APPLICATION OF THE DATA FOR NEA OBSERVATION

We can cover all the sky visible from the site within a few months, and therefore can contribute to international collaborative efforts for NEA detection. It is also important that follow-up observations be made to verify observations by our team and by United States teams. We therefore encourage observing astronomers and amateur astronomers to support this program by making those follow-up observations, which can be carried out using telescopes with only narrow field size (but requiring a minimum aperture of 0.5 m).

Our data handling method is shown in table 2. We will detect the NEAs in each field, using our automatic detection software developed specifically for detecting moving objects, within a day. Two weeks after its collection, we will release all raw data to the public via the Internet, after registering the data with the British Spaceguard Centre (BSGC). We can use our data for educational purposes between one day and two weeks after its collection.

Table 2. A time-scale for use of data collected by the telescopes.

Time after data collection	Activity
0-1 day	Detection of NEA and Space Debris
1-14 days	Educational Program
15-28 days	Make Data Available to the Public
28 days	Delete the data?

We are developing a new software product for this educational program. The system is very simple. One command requests transfer of data from our storage to the user's computer. For those lacking a sufficiently high-speed data transfer Internet line, data stored on a CD-Rom can also be used. Other commands activate a "blinking process" by which several images (usually five) alternate on the screen and "blink". The images are calibrated using fixed stars as points of reference. This enables one to detect moving objects easily (see figure 4) and to determine their positions (right ascension and declination) by just putting the cursor over each object. Then, either the determined positions are sent to the BSGC to determine whether each refers to a known or unknown asteroid, or one can download each asteroid's position from a home page operated by the Minor Planet Center. If an object has no known candidate asteroid, then it could be a newly discovered object. In all cases, we request the contributor to send all the position data of detected asteroids.

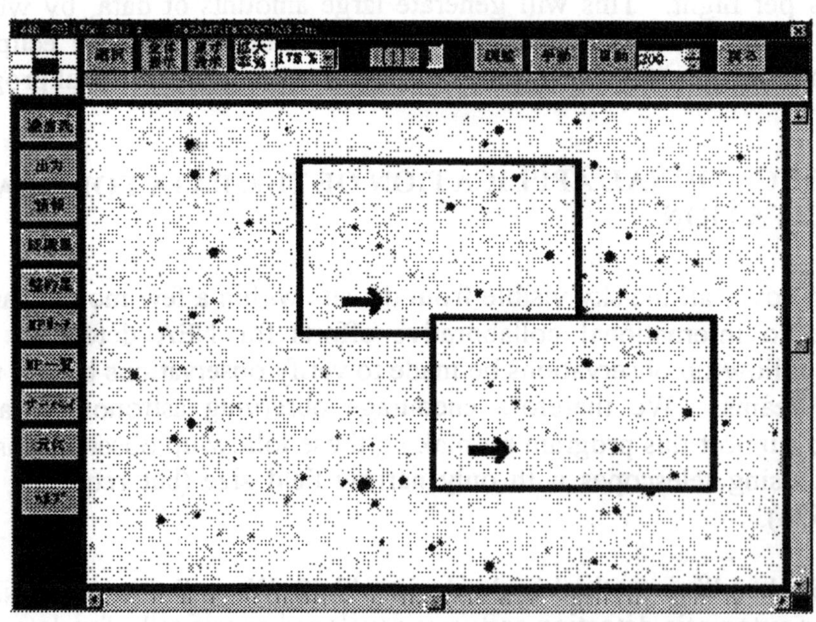

Figure 4. Displays of two frame images of an asteroid that is detected by the "blinking method" (the asteroid by the cursor is slightly to the right in the lower image).

Now, it is clear that schoolchildren and members of the public gain an understanding of how scientific measurements are carried out through this practical activity. One can also imagine how much they are excited by the discovery of new asteroid for which they will have the possibility of proposing a name to the International Astronomical Union after the asteroid's orbit is determined precisely. A few months later, after several follow-up observations have been carried out, they may also determine its orbital elements using a simple least square method.

The amount of our expected observational data is shown in table 3, taking into account weather conditions at the site. If we distribute to each group one field of 2 k x 4 k data obtained at five different times in a given night, there is high possibility that over 500 groups will obtain one set of data. This is enough for our international educational program, at least for its first period.

Table 3. Amount of observational data expected to be obtained by the one-meter telescope.

1 Exposure	10·2 k x 4 k CCD	→ 160 Mbytes
1 Night	Exposure time = 30 seconds	
	Read-out time = 8.3 seconds	
	Pointing time = 1.7 seconds	
		→120 Gbytes
1 Month	17 observable nights / month	
		→ ~2 Tbytes

4.   CONCLUSION

We, the Japan Spaceguard Association, are very keen to contribute to protection of humanity from the hazard of asteroid collisions by helping to detect all the hazardous large NEAs. Unfortunately, most people are not aware of the NEA problem. We have designed this new educational system not only to contribute to science education in general but also in an effort to efficiently publicize the problem.

The author expresses his sincere thanks to four staff members of the BSGC and to active members of the Japan Spaceguard Association for their intensive work to make this project successful. The educational software referred to was written in large part by Mr. Hiroshi Kaneda.

REFERENCE

Gehrels, T. 1994, Hazards due to comets and asteroids (The University of Arizona Press, Tucson).

NETWORKING OF SMALL ASTRONOMICAL TELESCOPE FACILITIES IN EDUCATION AND RESEARCH PROGRAMMES ON SUBJECTS SUCH AS VARIABLE STARS AND NEAR-EARTH OBJECTS[*]

S. Frandsen
Institute for Physics and Astronomy
Århus University
Bygning 520
DK 8000 Århus C, Denmark

ABSTRACT

The trend towards larger and larger telescopes has led to discussion about the future of the many smaller telescopes. Not all science can be carried out in a few nights on a giant telescope. At the same time, advances in automation and communication open up far more efficient and time-saving methods of observation, enabling very large datasets to be obtained and handled. Small telescopes allow a much larger community to participate in and carry out real science. Amateurs and students can become part of the scientific community and can contribute in a significant way in some areas of astrophysics.

In this paper, I will discuss modern studies of variable stars and briefly mention monitoring of the sky for Near Earth Objects and other special events.

1   PROJECTS INVOLVING NETWORKS OF SMALL TELESCOPES

Time is often an important factor when studying astronomical targets. Observations must be planned and a strategy implemented in order to obtain data at the right time and during the right periods. When observing oscillating starts with a rich spectrum of modes, daily side-lobes in the power spectrum of modes are very disturbing. Continuous viewing, either by networks or from space can remove or at least reduce these parasitic peaks. In other situations, one is looking for events and many hours are spent waiting for the right event to show up. It is a waste of a large telescope to have it run idle for long periods. This paper presents a number of examples of observing programmes involving networks of small telescopes.

---

[*] *This paper was presented at the "Ninth United Nations/European Space Agency Workshop on Basic Space Science", held from 27 to 30 June 2000 in Toulouse, France, and does not necessarily reflect the views of the United Nations.*

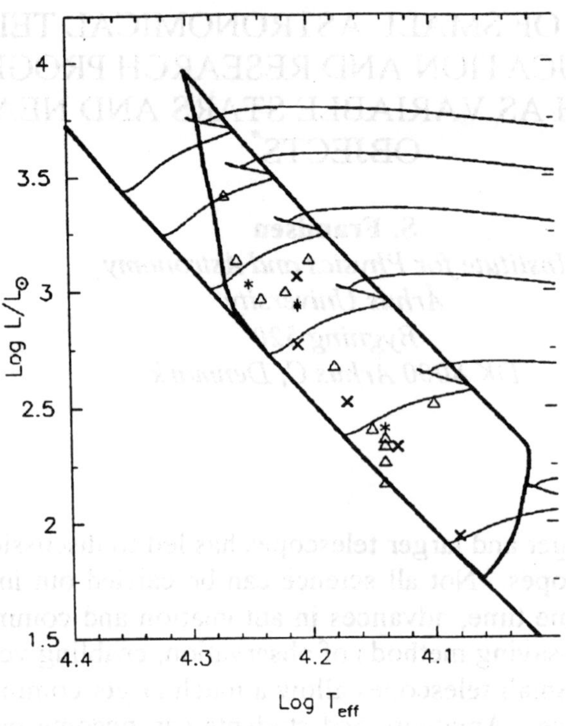

Figure 1: The location of the Slowly Pulsating B stars (SPB) instability region, with some real SPB's plotted in (from Aerts et al. 2000).

1.1 Oscillating stars

Variable stars present a rich source of information on stellar properties. Stars especially, with a rich set of non-radial oscillations excited, offer possibilities for using seismic techniques. Several classes of stars have been studied extensively. Large campaigns have been organized to disentangle the often complicated pulsation spectra. Among targets for such campaigns are white dwarfs, rapidly oscillating Ap stars (roAp's), δ Scuti stars and pulsating B subdwarfs. The stars have relatively short periods, up to a few hours, and mainly present pressure driven modes, with gravity modes appearing in the case of white dwarfs. Campaigns run for periods from one week to several months. Very long campaigns are needed even for the short period variables mentioned, because oscillations show up with very small frequency separation.

1.1.1 Slowly Pulsating B stars (SPB's)

An interesting class of stars, SPB's, is located slightly above the main sequence on the hot side of the instability strip (Fig. 1). The stars are slowly pulsating with periods in the range of 1 to 3 days, indicating that we are seeing gravity modes, non-radial pulsations driven by gravity. The instability is driven by the same opacity mechanism responsible for driving the β Cephei stars' oscillations with shorter periods (hours).

About 30 SPB's are brighter than the 6th magnitude (Aerts et al., 1999a, 1999b). The amplitude variations are quite small, only a few percent.

For professional astronomers, these stars present a problem as very long observing runs are needed. It is hard to find so much telescope time at normal observatories.

1.1.2 Subgiants

The Sun is known to oscillate in millions of different modes. The excitation comes from the random 'kicks' by convective elements, a process that also operates in other stars with convective envelopes. In solar-like stars, the oscillations are unobservable from the ground. The amplitudes are a few parts per million (ppm), which is totally out-of-range for small telescopes, and possibly also for large telescopes.

The amplitude increases considerably with luminosity and in the subgiant range the amplitudes are large enough and the periods short enough to change the picture. Oscillations have been seen by Buzasi et al. (1999) in α Ursae Majoris and by Edmonds and Gilliland (1996) in stars in the cluster 47 Tuc. For stars in the range G7 III to K0 III, amplitudes are expected in the range 100-200 ppm, and observing sessions should last 20 to 30 days. With 10 small telescopes and a duty cycle of 30%, the noise level obtained could be in the range 10-20 ppm. The signal-to-noise obtained would be S/N~10, good enough to permit detection of a small set of modes. A simulation of network observations for a δ Scuti star is presented by Kjeldsen (2000).

1.2 Near Earth Objects

A different strategy is used for the search and study of bodies very close to us. It is obvious that asteroids and comets, which pass close to the Earth, are of great interest. This is true, first, because of the chance, however small, that they would collide with the Earth, but also because the small distances involved permit detailed studies to be carried out. With small telescopes and charge coupled device (CCD) cameras, large fractions of the sky can be monitored and fast-moving objects located.

The largest effort at the moment in terms of surveying the sky is the Digital Sloan Sky Survey, which generates 200 Gbyte of data each night, although the Survey is directed towards extragalactic research. The Amateur Sky Survey (TASS) collaboration (Gombert and Droge, 1998, (http://www.tass-survey.org)) is aimed at chasing possible Earth-crossing asteroids, but is also useful for many other purposes.

Light curves for variable stars with large amplitudes can be monitored (e.g. semiregular variables). Gamma-ray burst afterglows can be monitored as well as other transient sources such as cataclysmic variables. 'Hot Jupiters' cause eclipses of a few percent, which would not be too difficult to detect were it not for the very small fraction of stars where this can be observed.

The major problem in this type of work is to organize the extremely large amount of data generated by CCD cameras. More ideas for targets for networks of small telescopes can be found by looking at the observing programmes of Automatic Photoelectric Telescopes (APT's) described later in the section on *Robotic Telescopes*.

2 TIME VERSUS SIZE

In science experiments, one constantly attempts to carry out more accurate measurements. A simple but costly way is to increase telescope size. Less costly, but difficult, is to improve instrumentation to increase efficiency and stability. A third option is to use more telescopes, which is straightforward in principle but has its own set of advantages and disadvantages.

Let us begin with one of the positive aspects. The scintillation decreases only slowly with telescope size. The noise level is given by

$$\sigma_{scint} = 0.09 D^{-2/3} \eta^{3/2} \exp(-h/8000m)\Delta t^{-1/2} \tag{1}$$

where D is the telescope diameter in cm, η the air mass, h the telescope elevation in m and Δt the integration time in sec. For a small telescope (D = 50cm) at sea level (airmass 1.5) and integration of t = 60s, the noise is σ_{scint} = 1.56 mmag. For bright stars using broad band filters, this is the dominating source of noise.

Now take 10 widely separated 50cm-telescopes. In this case, the combined noise from a one-minute integration is $\sigma_{10} = \sigma_{scint}/\sqrt{(10)}$ = 0.49 mmag. Compare this with the result obtained with a 2.5m-telescope, which is σ_{large} = 0.53 mmag. The 10 small telescopes do slightly better even though they have a collecting area that is 2.5 times smaller. For bright stars and broad filters, the scintillation noise dominates down to $m_v \sim$ 11, depending on the exact parameters of the instrument and target star.

The second advantage one can obtain is linked to the possibility of placing the 10 telescopes at different longitudes and obtaining measurements with less interruption. This is illustrated in the following simulation (Fig. 2) of the measurement of the light-curve of a δ Scuti star with 5 modes.

3 EXISTING NETWORKS

3.1 Photoelectric Networks

3.1.1 Whole Earth Telescope (WET)

The best known network is probably the Whole Earth Telescope (WET), which was created first to monitor light-curves of white dwarfs but later expanded to other targets as well. The web page is found a (http://bullwinkle.as.texas.edu). Around 15 sites participate, but only around ten sites are sufficient to provide good time coverage. Campaigns run for a couple of weeks.

3.1.2 Delta Scuti Network (DSN)

The Delta Scuti Network (DSN) mainly observes Delta Scuti stars, as indicated by the name of the network. The campaigns tend to become longer and longer. The activities are described on the Web page located at (http://www.deltascuti.net). The network has existed, as has WET, for some time, and extensive campaigns are carried out. About 15 sites are involved and a dozen δ Scuti stars have been observed.

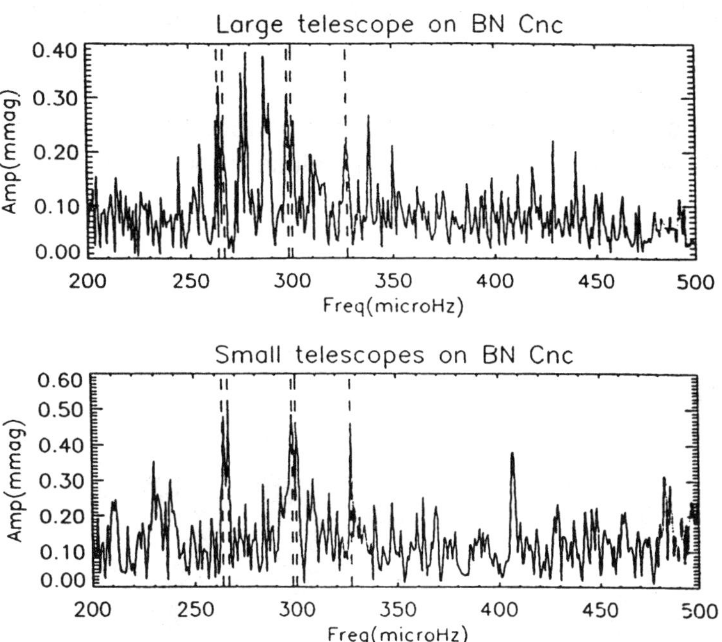

Figure 2: Amplitude spectrum of two weeks of measurements. A 50% duty cycle and some random weather statistics with a mean 60% of clear nights were assumed. The noise level was also assumed to be twice the scintillation noise. The first panel shows the results using a single large telescope, and the second shows the results using ten smaller telescopes. The oscillations are much easier to identify correctly in the second case.

3.1.3 STEPHI

A much smaller network uses identical photometers. The name is derived from Stellar Photometry International and is controlled from the Paris Observatory with only three observatories involved, in China, Tenerife and Mexico. The network is devoted to δ Scuti stars and has been active for some years. The Web page is located at (http://dasgal.obspm.fr/~stephi).

3.2 Differential CCD Photometry

With the progress in CCD detectors, small telescopes with a large CCD detector are no longer a remote dream. Such systems will probably replace photoelectric photometers at most sites due to their greater flexibility and higher efficiency. Some of the networks mentioned above are using CCD-based instruments already.

3.2.1 Small Telescope Array with CCD Cameras (STACC)

This network is more loosely organized and is directed to observations of multiple targets, such as stars in open clusters. Again, the main target is δ Scuti stars. A Web page describing its activities is located at (http://astro.ifa.au.dk/~srf/STACC).

As this is the area in which my institute has participated mostly, I will discuss this in detail. I will report on our experience with a large campaign on two δ Scuti stars in the Praesepe cluster (BN and BV Cnc), which are close enough together to be observed simultaneously.

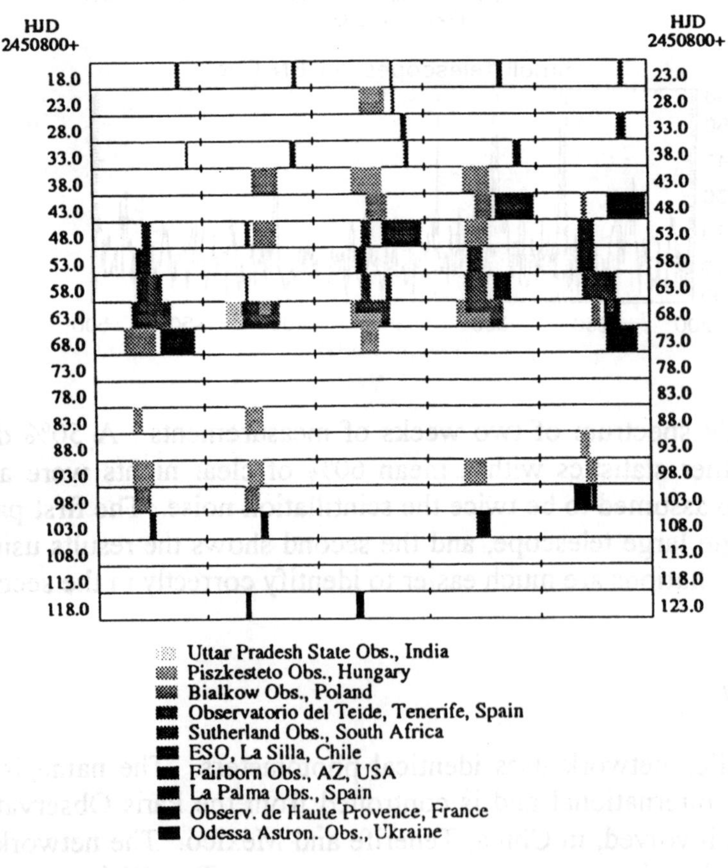

Figure 3: Distribution of photometric observations during the 1998 STACC campaign. Five consecutive nights are shown in each row (produced by A. Pigulski, Poland).

The campaign involved both a photometric aspect and a spectroscopic aspect. Figure 3 indicates the nights on which photometry was obtained.

The telescopes had diameters between 40cm and 1m. The results of the campaign were very different from site to site. This is indicated in the table, which gives r.m.s. values for the residuals after the light-curve has been subtracted. The final column gives the 'weight' each site has in the fit of the final light-curve for the star.

An illustration of the results can be seen in a light-curve from two of the nights, where data from three sites are present (Fig. 4). The point to be illustrated here is that when one relies on different groups at different sites participation on different levels, then a few 'best' sites will show up. Some sites have better instruments than others and some get more nights allocated and benefit from a better climate. The result is a quite inhomogenous dataset. The results from the STACC 98 campaign are nevertheless among the most precise for a δ Scuti star. And these are observations for two stars in one campaign!

A second result is the somewhat disappointing residuals that show up, which are more than a factor of two times the theoretical best. We obtained the best residuals around 3 mmag, and would expect the best around 1 mmag. This is partly explained by the nature of the star itself, which certainly is not matched by the light-curve used. It is also due to the difficulty of keeping the instrument response stable for periods of hours. At higher frequencies the noise decreases. A few sites suffered from severe drift problems. Finally, one site was testing a new camera.

Table 1. Residuals after fit of light-curve for data obtained at different sites.

Observatory	Residuals (mmag)	%
Tenerife	3.34	40.1
Piszkestetö	4.07	24.1
Fairborn		19.4
(v)[†]	3.99	13.7
(y)	5.90	5.7
Bialków	3.96	6.5
Sutherland		5.1
(v)[†]	4.20	1.8
(b)	3.43	2.1
(y)	4.53	1.2
ESO (Dutch)	5.22	2.7
La Palma	3.40	1.8
UPSO	8.44	0.2
Odessa	11.14	0.1
OHP	-	0.0

[†] Some sites were observing in more than one colour and the contribution is shown for each of the passbands as well as the total after converting to the V system.

3.2.2 ISTeC

Another network is the International Small Telescope Cooperative (ISTeC), for which the Web page is located at (http://www.astro.fit.edu/istec). It mainly provides a forum for organizing campaigns and does not have a scientific programme of its own. The list of members of the cooperative is quite long and gives a good indication of the location and capabilities of telescopes in the world. It is a good starting point for contacts to sites if a bright idea pops up and a network operation is required.

3.3 Time series spectroscopy

Light-curves can tell us about periodicities and, using different colours, something about mode identification. Time series spectroscopy provides access to a range of other techniques for measuring properties of oscillations. 50cm telescopes may be too small, but on 1.5m telescopes excellent results have been obtained, showing a richness of techniques for probing into details of stellar physics and the oscillations themselves. An illustrative example is the detailed study of the pulsation modes in the roAp star α cir by Baldry et al. (1998).

3.4 Political aspects

Networks are operated in many different ways. The most streamlined type consists of identical instruments under the control of a center. The solar network called the Global Oscillation Network Group (GONG) is an example of such a network. Everything was constructed and tested by the GONG team in Arizona and all data are shipped back from the stations. Data go through a pipeline reduction and are then made available for teams specializing in various different topics.

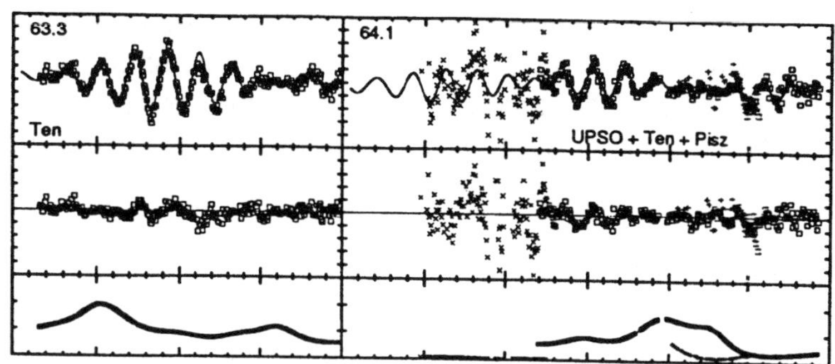

Figure 4: Observations from two nights in the middle of the 1998 STACC campaign. Upper panels show the light-curve, including a model light-curve. The middle panels show the residuals after subtracting this light-curve. The bottom panel gives the weights applied to the data points during the fit of the light-curve.

Another solar network is called International Research on the Interior of the Sun (IRIS), where the instrument was also constructed at a central site in Nice, France, but where the operations and data reduction is decentralized. The more active role of the people in the institutes connected with the observing sites has a large educational impact in small institutes and may open up opportunities for young people to join the international scientific community.

4 PRACTICAL CONSIDERATIONS

The success of a network depends on a number of issues – technical, economic, organizational and political – all in a complicated mix. The best networks operate according to a well-defined strategy.

4.1 Instrumentation

The possibilities with a network depend very much on the nature and location of equipment available. A modern telescope at a good site is a lot more productive than an old telescope in, for instance, Europe, where often only 10-20% of nights allocated are useful.

Modern detectors also make a large difference. If one can afford a large field-of-view CCD camera with special software for readout and data reduction, giving light-curves in real-time, then observations can be carried out and good data obtained even in poor weather and close to city lights. This means that campus observatories can contribute.

Spectroscopy is more 'expensive' in all respects. But as high resolution is not necessarily needed, a simple, cheap instrument can be constructed. The most expensive item is still the detector (the CCD camera). Data reduction is probably the most critical issue. Dedicated software must be written that extracts the spectroscopic quantities from the images.

4.2 Storage and communication

The progress in computing and networking makes the storage and communication easy in case of photoelectric photometry, where the volume is very small. In the case of differential CCD photometry, one should strive to develop real-time reduction software. Then, the amount of storage and data transfer is reduced and becomes similar to the output of a classical photometer. If images have to be stored or transferred to remote sites, then the volume is a problem.

For monitoring programmes requiring a large number of CCD images to be compared and checked together, the organization of this part of the programme is the key to success.

4.3 Data reduction and manpower

To run a network efficiently, a fair amount of funding and manpower must be available. Managing the operation of the network and the data collected during campaigns is time-consuming and expensive. There is often a considerable delay in the output of the science part after data has been collected. The data quality often varies between different sites and a lot of experience is needed before the data can be combined and results extracted that are as accurate as possible.

4.4 Quality control

The ideal situation is to have a network where there is good feedback to the participating teams. Few teams obtain perfect data the first or second time around.

5 ROBOTIC TELESCOPES

Long-term observations of long period variables or the countless hours spent waiting for some spectacular event are often extremely tedious. Only the hope that it will lead to an exciting discovery makes you carry on. In addition, the instrument is often located far from home or the office, so that time spent observing is often not used very efficiently.

A very attractive solution is to employ robotic telescopes and observe from home in an armchair or from the office, if the telescope is placed conveniently in longitude on another continent.

Such facilities are becoming increasingly popular, also driven by the option of using them for public access to the sky. The role of such telescopes in education will probably explode soon. An example is the Iowa Robotic Telescope Facilities (IRTF). The IRTF Web page is located at (http://denali.physics.uiowa.edu), where you can also find links to some other facilities.

The Institute for Astronomy at the University of Vienna in Vienna, Austria, employs two remote telescopes, with the nicknames 'Wolfgang' and 'Amadeus', placed at the Fairborn Observatory in Arizona, which are used to monitor stellar activity. The Web page (http://www.astro.univie.ac.at/~kgs/APT) also has links to other sites.

6 CONCLUSION

Even in these days of very large telescopes, there is still a lot of exciting science to be carried out with small telescopes. With better detectors and more automation, these telescopes can participate in many important projects even in the future. Countries that do not have the possibility to join the league of nations with access to large telescopes can join networks and obtain contacts that would give them access to the larger world of astrophysics. They also serve as an excellent platform for students to start learning experimental astronomy, and for giving the general public the opportunity to conduct astronomical research.

REFERENCES

Aerts, C., De Cat, P., Peeters, E. et al. 1999a, *A&A 343*, 872

Aerts, C., Uytterhoeven, K., De Ridder, J. et al. 1999b, *The Third MONS Workshop: Science Preparation and Target Selection, ed. T.C. Teixeira and T.R. Bedding*, 59

Aerts, C., De Cat, P., De Ridder, J. et al. 2000, in *The Impact of Large-Scale Surveys on pulsating Star Research, ed. L. Szabados and D.W. Kurtz, ASP Conf. Ser. 203*, 395

Baldry, I.K., Bedding, T.R., Viskum, M., Kjeldsen, H. & Frandsen, S. 1998, *MNRAS 295*, 33

Buzasi, D.L., Catanzarite, J., Laher, R., Conrow, T., Shupe, D., Gautier III, T.N. and Kreidl, T. 2000, *Astroph.J. 532*, L133

Edmonds, P.D. and Gilliland, R.L. 1996, *Astroph.J. 464*, L157

Gombert, G. and Droege, T. 1998, *Sky & Telescope*, Feb., p. 43

Kjeldsen, H. 2000, in *The Impact of Large-Scale Surveys on pulsating Star Research, ed. L. Szabados and D.W. Kurtz, ASP Conf. Ser. 203*, 415

THE NETWORK OF ORIENTAL ROBOTIC TELESCOPES (NORT): AN ARAB PROJECT[*]

Roger Hajjar
Department of Sciences, Notre-Dame University, Lebanon

François Querci and Monique Querci
Observatoire Midi-Pyrénnée, Toulouse, France

Hamid Al-Naimiy
Institute for Astronomy and Space Sciences, Al Al-Bayt University, Jordan

and

Khalil Konsul
Jordanian Astronomical Society, Jordan

ABSTRACT

In this paper, we present the Network of Oriental Robotic Telescopes (NORT) as an Arab project. Arab countries cover a large part of the longitude range of the initial NORT project proposed by the Quercis. The Network's adoption by a number of Arab astronomers and the Arab Union for Astronomy and Space Sciences (AUASS) offers a solid basis for the rebirth of Arab astronomy, astrophysics and space sciences. The number of national projects for telescopes in several Arab countries makes the NORT a framework of choice to coordinate and stimulate these separate efforts by the creation of a NORT steering committee.

1 INTRODUCTION

The Network of Oriental Robotic Telescopes was initially proposed by the Quercis in 1993 (Querci & Querci 1993). The Network is to be devoted to the study of variable stellar phenomena by ensuring a long term 100% coverage of the variability, thus allowing for a complete analysis of periodic and non-periodic changes in stars. This is essential for studies of astroseismology, for example. The Network would cover longitudes from Mauritania in the West to China in the East, thus effectively complementing the American and Pacific longitudes.

The NORT would network a series of small to medium sized telescopes (1m to 2m) located at or near the Tropic of Cancer, covering the entire longitudinal range previously

---

[*] *This paper was presented at the "Ninth United Nations/European Space Agency Workshop on Basic Space Science", held from 27 to 30 June 2000 in Toulouse, France, and does not necessarily reflect the views of the United Nations.*

mentioned. This region is interesting because it includes areas of high mountains in semi-deserts, which are very good candidate sites with excellent seeing conditions and a large proportion of clear nights. The interest in NORT within the Arab World stems from the fact that Arab countries cover a huge chunk of the longitudinal range that the project would like to cover.

2   THE NORT

2.1   Objectives

The major aims of the project enter into the realm of what the United Nations Office for Outer Space Affairs has set as a primary mission: to make basic space science accessible to developing countries. The NORT fixed the following goals for its mission:

a.   Enhance research in astronomy and astrophysics in developing countries through participation in variability studies (photometric and spectroscopic) of stellar objects. Such studies require complete time-coverage of the light curve of an object and long runs on telescopes and are therefore better carried out with small to medium sized telescopes. This keeps costs low!

b.   Provide, through these studies, a basis for participation in international endeavours and observations on large telescopes and interferometers.

c.   Permit, through astronomy, the development of new, cutting-edge technologies, which would also be of relevance to other fields. This would ultimately lead to an improvement in the contribution of Arab countries to large space science projects.

d.   Develop national observatories, an activity that would also support teaching and training of astronomers attending national universities.

e.   Offer summer courses taught by Arab and French astronomers in France and other countries, for students and/or amateurs, in order to improve education in astronomy and astrophysics.

The NORT project was selected in 1996 as a follow-up project to the United Nations/European Space Agency workshops on basic space science (document A/AC 105/657, 13 Dec 1996) and recognized by the AUASS during the second Arab Astronomical Conference held in Amman in 1998.

2.2   NORT research scientists

NORT currently groups a number of scientists from various countries in which some interest in developing basic science is emerging or in which astronomy and astrophysics are established to a certain degree. The following list is not exhaustive and may include more research scientists in the future:

1.   Pr. François R. & Dr. Monique Querci, Observatoire Midi-Pyrénées, France
2.   Dr. Ph. Stee & Dr. F. Vakili, Centre d'Etudes et de Recherches Géodynamiques et Astronomiques (CERGA), France
3.   Pr. H. Al-Naimiy, K. Konsul & Mr. H. Sabat, Jordan
4.   Pr. M. El Eid, Dr. R. Hajjar , Dr. K. Hussein & Mr. J. Bittar, Lebanon

5. Eng. H. Gashut, Libyan Centre for Remote Sensing and Space Science (LCRSSS), Libya
6. Pr. S. Kadiri & Pr. H. Touma, Morocco
7. Dr. A. K. Hamdo & N. Chamsi, Syria
8. Dr. N.E. Jaidane & Dr H. Atteb, Tunisia
9. Pr. A-H. Al-Khersi, Yemen

The groups mentioned operate within the longitudinal range of the Arab World and the Mediterranean Basin. From the adjacent area from Iran to the Far East, two groups have manifested interest in NORT:
1. Pr. S. Z. Farooqui, Karachi, Pakistan
2. Dr. M. R. Hidayat, Kuala Lumpur, Malaysia

It is clear, then, that the NORT has generated a lot of interest in Arab countries, which is the basis for the proposal to establish, as a subgroup, an Arab Project for NORT.

3 ARAB PARTICIPATION IN NORT

3.1 Current Status of Astronomy and Astrophysics in the Arab World

The last few years have seen a number of initiatives develop in various Arab countries based on the importance of basic space sciences in the modern world and as an effort towards restoring astronomy in the tradition of the Old Arab Empire. Some groups are now working on various projects of interest to them. We have cited below a few of the research efforts and telescope projects underway.

Research interests cover a broad range of topics, from atmospheric physics, often related to studies of turbulence and seeing conditions, to varying interests in stellar astronomy and astrophysics. These include:
a. Algeria – Seghouani team – atmospheric parameters of relevance to astronomy;
b. Egypt – Hamdy group – early type star variables (photometry);
c. Jordan – Al-Naimiy/Sabat team – binary systems and X-Ray binaries;
d. United Arab Emirates – Gessoum team – thermonuclear reactions of light nuclei in astrophysics;
e. Lebanon – Hajjar – circumstellar envelopes of Ae/Be and young stars, imaging polarimetry; El -Eid team – stellar evolution
f. Lebanon/France – Bittar/Stee – Be and B[e] stars, photometry, spectroscopy and interferometry
g. Saudi Arabia – Al-Malki group – large envelope (polarization), RSCVn stars
h. Syria – Al Mousli team – remote sensing and research into sites using satellites such as Meteosat and Spot;
i. Bahrain – Dalal – space physics, the Earth's magnetosphere, theoretical astrophysics.

Furthermore, one should note the presence of a number of institutes of astronomy and of other fields relating to space.

Observatory projects and operating observatories are increasing in number in various nations. Collaborative efforts are now underway through various initiatives, some under the auspices of the AUASS. Once again, the following is non-exhaustive list reflecting information gathered by the authors:

a.   Egypt – 1.93-m existing telescope;
b.   Jordan – 40-cm existing telescope and a proposal currently under study for a 1.5-m telescope;
c.   Kuwait – 1.5-m telescope project;
d.   Lebanon – Preliminary study towards a national observatory (1-m telescope);
e.   Morocco – 55-cm existing telescope;
f.   Syria – meteorological mapping already completed for a still undefined observatory project.
g.   Tunisia – 1-m telescope project;
h.   Saudi Arabia – a number of existing 60-cm telescopes;
i.   Libya – discussion to explore the feasibility of and aims for a national observatory (1-m or 2-m class telescope).

3.2   The NORT factor

The contribution of NORT to this situation is very important. It can provide a framework for coordination and collaboration in achieving all the steps required for the implementation of observatories in the different Arab countries. Furthermore, standards developed within the NORT will produce an opportunity for collaboration that will enhance the capacity of participating countries. The number of projects achievable with networks covers a wide spectrum of technologies and techniques to produce through the unified infrastructure of NORT a rapid growth in basic space sciences. To list a few of the topics of study:

- Intrinsic variables such as low-mass red giants, dust-shell stars in post-AGB phase (bipolar flow) and RV Tau stars;
- Eruptive variables such as Be stars and RCB stars;
- Binary stars, mainly with matter exchanges;
- Meteor and comet searches, behavior of comet coma (condensation, tail, flares, jets);
- Near-Earth objects, including discovery and follow-up of some those with a high angular speed (around 2 or 3 degrees by day);
- Search for planets around nearby stars;
- Surveys, for instance for Be stars, clustering in star-forming regions and polarimetric surveys.

The existence of such a large network in areas with more then 200 to 250 clear nights per year will undoubtedly produce a large body of data making a suitable contribution to the Virtual Observatory project.

NORT can therefore contribute significantly to capacity building in basic space sciences in the Arab World and other participating countries such as Pakistan and Malaysia.

4   STEPS TOWARDS NORT

It has been a few years now since the inception of the idea of a NORT. Querci & Querci initially produced meteorological maps of the region of interest for the project showing the potential of region near the Tropic of Cancer. They used a grid of 250 x 250 km. It is obvious that the current interest in the Network should lead to a number of important steps at the level of participating countries while benefiting at the same time from expertise already available among partners. A basic plan looks like the following –

a.   Using archives of meteorological satellites such as Meteosat and NOAA:
- Locate sites with very good weather conditions for astronomy from Morocco to Tabla-Makan; and
- Choose sites not submitted to the same weather patterns.

b.   In-situ studies of selected sites:
- Analyze the quality of the atmosphere, including factors such as opacity, seeing conditions, sky brightness (GSM monitors, Nice and Marrakech laboratories, or DIMM);
- Analyze the level of light pollution or electromagnetic pollution;
- Analyze the quality of facilities, including road access, electricity and water.

c.   Create a NORT Steering Committee to coordinate and overview efforts towards implementation of the Network. The Committee would also serve as a scientific advisory committee for the project.

5   CONCLUSION

The aims of the NORT can be summarized as follows:
- To promote education and training in universities with 1-m class telescopes, towards the rapid participation of Arab countries and other affiliated partners in international astrophysics.
- To contribute to a complete time coverage and follow-up of variable astronomical objects with 2-m class robotic telescopes and in areas inaccessible to large specialized telescopes.
- To promote a new type of cooperation with larger facilities such as optical and infra-red long-baseline interferometers, using facilities such as GI2T & GI3T, ISI, VLTI and Keck I and II, which are mainly required at critical phases of variability of the objects under study. These phases are detected through permanent follow-up by networks such as NORT.

The Adoption of the NORT project by the AUASS and the strong interest in Arab countries requires the following steps to be taken:

- Coordination of efforts through the AUASS, through the creation of an Arab Steering Committee for NORT.
- Active collaboration with international organizations such as the United Nations, the International Astronomical Union (IAU) and the European Space Agency (ESA) to implement the project in the short term.

ASTRONOMY IN LEBANON: PAST, PRESENT AND FUTURE PROSPECTS[*]

Roger Hajjar
Department of Sciences, Notre-Dame University
Po.Box 72 Zouk Mikael, Zouk Mosbeh, Lebanon
E-mail: rhajjar@ndu.edu.lb

ABSTRACT

This paper reviews briefly the recent history of astronomy in Lebanon through the work carried out mainly at the Ksara and Lee Observatories. The current situation is presented. The number of astronomers is still very low but interest is growing leading to a project for a national observatory. A brief review of the report that will be submitted to the appropriate authorities is included, which discusses possible sites and the selection process.

1   INTRODUCTION

Since the early nineties, an observatory project has been under discussion among the scientific community in Lebanon sparked by the initial input of Dr. François Querci, aiming at the development of the Network of Oriental Robotic Telescopes (NORT) project. There were no astronomers or astrophysicists working at that time in Lebanon and the initiative originally came from some physicists and the Dean of the Faculty of Sciences at the Lebanese University (LU). The project remained in a state of limbo until the last half of the nineties when some astronomers and astrophysicists started working in the Lebanese universities, and the initial effort produced a Ph.D candidate now completing his thesis in France. In present-day conditions, astronomy in Lebanon is reviving. It started with separate efforts in several universities. Concerted and coordinated action is now underway to reestablish an inter-university center for astronomy and space sciences.

In view of this burst of activities, I will briefly review the recent history of astronomy in Lebanon (20th century) then paint a picture of the current status of astronomy in the country, in the fields of education and research. The last part of this paper will be devoted to examining the future of astronomy, sketching the project for implementation of a national observatory.

---

[*] *This paper was presented at the "Ninth United Nations/European Space Agency Workshop on Basic Space Science", held from 27 to 30 June 2000 in Toulouse, France, and does not necessarily reflect the views of the United Nations.*

2   PAST

The recent history of astronomy in Lebanon is marked by the presence of two observatories: The Lee Observatory founded by the American University of Beirut (AUB), then called the Syrian Protestant College, and the Ksara Observtary, owned and operated by the Jesuits in the Bekaa Valley in the eastern part of the country. The Lee Observatory was and still is located on campus of AUB in the heart of Beirut. It should pointed out that, for different reasons, both observatories are no longer in operation. The dome of the Lee Observatory at AUB is still present but the instrument has lost its lenses and needs a major rehabilitation to be operational. It is mainly of historical significance. The Ksara Observatory has been sold with the wine making industry (Ksara) to private interests. The dome can still be seen but all instruments were dismantled a long time ago.

2.1   The Lee Observatory

The Lee Observatory is linked to the beginnings of AUB. Soon after the inception of the University in 1866, a course in astronomy was initiated and taught to all college students. It remained compulsory for all senior students until 1912. Efforts to build an observatory were intitiated by Van Dyck in 1874. The first equipment purchased was a 10-inch Newtonian, an equatorial clock and a prime-meridian transit, as well as some meteorological equipment. A 12-inch refractor was bought later, when the Observatory was rebuilt in 1892. Instruments that were successively acquired included a spectroscope, a 7-inch photographic doublet (1906) and a spectrohelioscope (1927). The instruments were too heavy for the mount and ended up wearing the gears.

The Observatory was used for time-keeping and for religious purposes such as determining the start of Islamic months. Research interests varied depending on the director of the Observatory and the instruments available. Some observations of the Halley comet were carried out by Joy in 1910, as well as observations of variable stars and double stars. Students recorded solar activity. The Observatory had been equipped since 1900 with a seismograph, leading to work in that field.

Astronomy courses stopped in 1947 and were revived again in 1955 thanks to the efforts of Owen Gingerich who worked on the rehabilitation on the Observatory. Gingerich remained for three years as the director of the Observatory. He was succeeded by Frans Bruins who was the last person to head the Observatory. All activities were ceased in 1979.

2.2   The Ksara Observatory

The Ksara Observatory was founded in the early 1900's (1906 or 1907). It was equipped with an 8-inch refractor. The instruments used included a spectroscope and a microphotometer.

The Observatory was mainly used for meteorology and seismology. Some serious efforts in astronomy were made by Rev. Fr. Plassard, the last director of the Observatory. He remained in charge of the Observatory from the late 1940's until 1979. His main interest was in the detection of telluric bands. This led him to the development of the microphotometer. Other interests were in meteorology and gravimetry. He produced a gravimetric map of Lebanon as well as a climatic atlas of the country. Early on, the Observatory was also used for variable star observations. All activities were also ceased in 1979.

3 PRESENT

3.1 Education

A major review of the school curriculum was completed in 1997 by the Lebanese government, leading to the introduction of topics in astronomy at different levels of primary and secondary education. In primary education, students should learn about day and night, the four seasons, the apparent movement of the Sun, shadows, times of day, phases of the Moon, the Solar System, the Earth's movements and their consequences, the rotation of Moon and satellites. In secondary education, the topics encountered are: historical development of astronomy; the Solar System; the evolution and dimensions of the Universe; instruments for observation, including telescopes, radiotelescopes, space stations and satellites; cosmology: the Big Bang, age of the Universe, Hubble's law, and the expansion of the Universe; and the life and death of stars.

In post-secondary education, the situation is different. Almost no astronomy is proposed for study by university students. AUB has been offering a non-calculus-based course in astronomy for the last few years, and Notre-Dame University (NDU) is introducing such a course starting in spring 2001. It should pointed out that a B.Sc. in Physics is offered at the LU, AUB, NDU, Université St-Joseph, and the Beirut Arab University (BAU). AUB and LU offer graduate programs in physics. At AUB, a student may complete a thesis in astrophysics since two astrophysicists are present among faculty members.

3.2 Astronomers, Astrophysicists, and Research

Three astronomers are currently based in Lebanon, and a student on a Ph.D scholarship from the Lebanese University will soon complete his thesis in France and is expected to join the group by Fall 2001.

At AUB, Mounib El-Eid works on stellar evolution and Jihad Touma has an interest in dynamical systems applied to astrophysical situations. At NDU, Roger Hajjar works on star formation. Jamal Bittar, the Ph.D candidate, is working on the circumstellar environment of Be stars. Recently, the number of Ph.D candidates has increased, who are working on several different topics and are expected to increase the number of astronomers in the country in the next three to four years. To this group, one must add some physicists with an interest in the development of astronomy in Lebanon.

It is obvious that the size of the "workforce" in astronomy, per capita, is still small compared to other countries. Nevertheless, Lebanon may build a "critical mass" quickly if a national project develops fast enough.

3.3 Amateurs

The number of amateurs in Lebanon is not well known because of the lack of a support organization. Recently, a number of clubs have seen the light of day. One club is affiliated to the BAU and another to NDU. At the LU northern branch, some physicists are initiating popular activities in astronomy. In 1999, a private company, Al-Kawn, started to organize astronomy summer camps in various parts of the country, and another company is promoting astronomy in school through the use of an inflatable StarLab planetarium.

4 FUTURE PROSPECTS: THE ROAD TO A NATIONAL OBSERVATORY

Astronomy in Lebanon is on the rise. This is mainly due to efforts of individuals working separately. More and more, these people are meeting in a concerted effort to increase the Lebanese potential in astronomy and basic space sciences. Presently, a number of Lebanese scientists from different universities have met to discuss the the need and required steps to establish an inter-university national observatory.

As a first step, a report has been drafted to propose a plan for the advancement of astronomy in Lebanon and pave the road to a national observatory.

In the field of education, emphasis has been put on post-secondary education through the implementation of minimal requirements in astronomy and astrophysics for physics majors. The aim is to give a foundation in space sciences to high-school teachers who are going to educate future university students in science. In college-level courses, there are plans to introduce basic observational projects to train students in some of the recent techniques of observation in astronomy. Furthermore, there is a need to raise the level of the amateur community in Lebanon through the introduction of technology and the increase and deepening of their knowledge of astronomy.

4.1 The Road to a National Observatory

In this project and the plan proposed for it, we tried to address the needs of research and education. Some well known steps need to be achieved for the implementation of such a project, the most obvious being the selection of a site for the actual buildings and dome. Inter-university cooperation on major projects is just starting in the country, and we chose to use the requirements of site selection to bring about the necessary atmosphere and skills towards the achievement of the project.

We opted for a 1m-class telescope. A 1m telescope opens the way to a larger number of projects then smaller telescopes. It is still a small observatory but is somewhat larger than a "traditional" amateur observatory. Furthermore, it can reach fainter objects than a 40 or 60 cm telescope and thus provides for a larger sample of objects in a given field of study.

Lebanon is blessed with a lot of moutains with a large number of summits above 2000 m. Accessibility playing a major role in a training and teaching observatory, we initially selected four locations for a possible observatory, each of which should be studied for light pollution, transparency, turbulence (seeing) and weather conditions. We plan to put each site under the responsibility of one university and to proceed as follows:

1. Equip each site with: with a **30-cm telescope, CCD camera UBVRI filters,** a **computer,** a **Plexi dome** observatory and housing (a small cabin with WC and a cell phone). A **weather station** would maintain logs of wind speed and direction, humidity, dew point and logs of observers (percentage clear time and clouds). At the end of the test period, the equipment would remain the property of the University Astronomy Club or other appropriate body.

2. Observing Procedure: **Common list of bright standard stars** to observe (m_V<8 or 9). Sites are to be **monitored on the same dates, looking at the same stars** for a specific date. Sky conditions and weather data logs are to be noted for each night, clear or not. **Charge coupled device (CCD) data format and headers should be identical for all sites.** There should be **centralized data processing** in a unified center.

3. Manpower should be supervised by a **PI in each university.** Work should be carried out by **university junior or senior students** along with the PI, or **trained astronomy club members,** divided in **teams of two,** with **rotation between teams** in the Observatory. This work is to be **credited to students** with a) credit towards their degrees, or b) credit towards their senior projects, **AND** c) authorship on publications.

4. Turbulence (seeing conditions) should be monitored using GSM, DIMM, or other techniques.

A GSM or a DIMM could be duplicated but would increase the costs. If a single intrument is used, comparing the data might lead to problems. One possible solution might be to use techniques for daylight seeing monitoring (limb of the sun) to night observing (moon).

The advantages of using this standardized approach are obvious. Equipment and observational procedures are common to all sites and data reduction is centralized. This produces a common zero point and calibration and easily comparable data. Political bickering will be then reduced to a very very low level!

The spin-offs of such a plan are also tremendous:

1. Small observatories for universities and astronomy clubs.

2. Student training in observational techniques.

3. Amateurs prepared to contribute to hard science!

4. Interest generated = graduate students in astronomy and astrophysics.

5   CONCLUSION

The burst of activities in astronomy in Lebanon and the presence of a nucleus of astronomers and astrophysicists, as well as other interested scientists, in Lebanese universities provides a good basis for the advancement of basic space science in the country. This is materializing through the project to develop a national observatory.

Some of the lessons learned in the last few years show that astronomy needs to be developed at all levels, from elementary schools to universities, amateur and professional alike. Astronomy is fun and we ought to show it!

Acknowledgments: The author would like to thank Dr. Mounib El-Eid, chair of the department of physics at AUB, who was kind enough to provide me with a document written by Dr. Frans Bruins, the last director of the Lee Observatory, on the history of the Observatory. Thanks also to Rev. Fr. Jacques Plassard, the last director of the Ksara observatory who provided me with a wealth of details about the work done in the Observatory during his long years of service. The author acknowledges the contribution of Dr. Hans Haubold of the United Nations Office for Outer Space Affairs through whom support whose granted for participation at the 9th United Nations/European Space Agency Workshop on Basic Space Science. And most importantly, Dr. François Querci, who provides continued support and drive for astronomy in developing nations and especially in Lebanon.

ASTRONOMY AND SPACE SCIENCES IN JORDAN[*]

Hamid M. K. Al-Naimiy
Institute of Astronomy and Space Sciences
Al al-Bayt University, Mafraq, Jordan

Khalil Konsul
Arab Union for Astronomy and Space Sciences & Jordanian Astronomical Society
Amman, Jordan

ABSTRACT

The aim of this paper is to summarize the activities and research projects in astronomy and space sciences (AASS) in the following Jordanian organizations and institutions:

1.   Jordanian Astronomical Society (JAS)

•   Popularization of astronomy and space sciences as an aid for education and development, for instance in universities, schools, the media (educational television, radio programs and hotlines), planetariums and public lectures
•   Observing programs in Al-Azraq camp for astronomical events such as comets, meteor showers, and lunar and solar eclipses.

2.   Al al-Bayt University

    a.   The Institute of Astronomy and Space Sciences (IAASS) has M.Sc. research projects and curricula in: Astronomy and Astrophysics, Space Sciences, Remote Sensing (Science and Technology), Environment and Water Resources

    b.   Maragha Astronomical Observatory (MAO) has a 40cm Meade optical telescope with CCD Camera

3.   Arab Union for Astronomy and Space Sciences (AUASS)

4.   Islamic Crescent Observational Program (ICOP), supervised by a Higher Committee from AUASS and JAS.

The paper also summarizes other activities in some Jordanian organizations and discusses future expectations for astronomy and space sciences in Jordan.

---

[*] *This paper was presented at the "Ninth United Nations/European Space Agency Workshop on Basic Space Science", held from 27 to 30 June 2000 in Toulouse, France, and does not necessarily reflect the views of the United Nations*

INTRODUCTION

Most of us know that AASS are important fields of research, knowledge and culture. They are the origin of both eastern and western sciences. Astronomy is an aspect of humanity's attempt to study and understand celestial phenomena, part of its never-ending urge to discover the order of nature. The science of astronomy was well advanced in ancient Mesopotamia, Egypt, Greece, India and China. Over the past 4,000 years, astronomy has been intertwined in many aspects of our culture, has stimulated the growth of science and technology and, in a continuous interplay with religions, has substantially influenced history.

In all Muslim countries astronomy has been treated as a basic science and applied successfully in the Islamic Mawaqeets, which determine of the Islamic praying times, fix the beginning of the Hijra months in general and Ramadan, Shawal and the Al-Hija in particular and calculate the Qibla direction. In this manner, astronomy and space sciences were established in different Arab countries, including Jordan.

The Hashemite Kingdom of Jordan is situated in the northwest corner of the Arabian Peninsula, with a population of around five million people and an area of about 570,000 km^2. Jordan features a highly diverse topography, from desert areas in the east and south-east, to hilly areas in the north and south, down to the Jordan Rift Valley, which includes the Dead Sea, the lowest point on the Earth's surface, averaging 400 m below sea level. The capital city of Amman, which lies in Jordan's mountainous north-central region, enjoys moderate, Mediterranean-like climate, making it ideal for astronomical observations.

Modern astronomy started in Jordan in September 1987, when the Jordanian Astronomical Society (JAS) was founded in Amman. The Institute for AASS in Al al-Bayt University (IAASS) was founded in September 1994. In September 1998, the Arab Union for AASS was established. There is now a plan to establish in Jordan the regional Centre for Space Science and Technology Education for Western Asia (Al-Naimiy, 2000).

The aim of this paper is to describe the development of AASS in Jordan over the last 15 years and to discuss future expectations for the field in Jordan.

JORDANIAN ASTRONOMICAL SOCIETY (JAS)

JAS, formerly known as the Jordanian Amateur Astronomers Society, is the only organization in Jordan that popularizes amateur astronomy. JAS was founded in Amman in September 1987 and is located in Haya Cultural Center. This society has promoted astronomy and space sciences not only throughout Jordan but also in other parts of the

Arab world. JAS currently has about 300 members, 100 of whom are active in club programs. (http://www.jas.org.jo) (Ohda, 1996).

Objectives

- Assembling amateur astronomers in Jordan and in the Arab World, in order to develop their astronomical hobby through the exchange of astronomical data, information, articles, and observational expertise, as well the provision of services for amateur astronomy, to the extent possible, such as astronomical equipment, periodicals, books, references, hotlines and planetariums

- Popularization of astronomy, spreading awareness about astronomy among members of the public, and ascertaining the importance of astronomy and applied sciences in everyday life, and their interrelationship with other fields of basic science and technology

- Transferring astronomy in Jordan and the Arab World from the amateur to the professional level, and from individual initiatives to institutional frameworks

- Assisting the teaching of astronomy in schools and universities

Ways And Means of Fulfilling JAS's Objectives:

- Weekly lectures at Haya Cultural Center

- Publications, such as "Pleiades" and "Al-Debran", the monthly internal publication of JAS

- Publishing astronomical articles and essays in local newspapers and the mass media

- Distributing astronomical periodicals (such as "Astronomy", "Sky & Telescope", "Astronomy Now", "Popular Astronomy", and "The Planetary Report"), as well as material such as astronomical books, video-tapes, posters and astronomical software

- Weekly and monthly observation of the sky and of major celestial events, using the naked eye and astronomical equipment (such as binoculars and telescopes)

- Organizing "Astronomical Camps" in the Jordanian deserts, which are usually devoted to certain astronomical events, such as meteor showers or comets (http://www.jas.org.jo/camp.html)

- Cooperation with other institutions, Jordanian, Arab or international, with similar activities; for example, JAS is a member of the International Meteor Organization (IMO), the International Occultation Timing Association (IOTA), the American Meteor Society (AMS) and the Royal Astronomical Society of Canada (RASC). (http://www.jas.org/out.html)

- Delivering lectures and seminars and organizing "Star Nights" at various universities, schools, clubs, societies and youth camps

- Founding "Astronomy Clubs" in various schools and universities, which are continuously supervised by the Society (http://www.jas.org.jo/clubs.html)

- Organizing "Astrofests", activities dedicated to schools, which may include an opening ceremony, astronomical lectures, an astronomical exposition, and a Star Night

- Organizing "Scientific Astronomical Days", in cooperation with universities and other institutions such as professional unions

- Organizing conferences about astronomy and space sciences, usually in cooperation with universities and other agencies (http://www.jas.org.jo/confe.html). JAS was especially involved in organizing the 1st (1992), 2nd (1995), 3rd (1998) and the 4th Arab Conferences in AASS. These were organized in cooperation with universities, particularly Al al-Bayt University. The outcome of 3rd conference was the establishment of AUASS. AUASS, JAS and IAASS are now preparing to organize the 4th Arab Conference in AASS at Al al-Bayt University between 28 and 30 August 2000.

- Organizing special training courses on various topics in astronomy, both for students and teachers, some of which are held in cooperation with the Ministry of Education

Other JAS activities

Since its inception, JAS has focussed especially on the visual observation of meteors, comets, planets and the crescent fixing the beginning of Hijra month. More than 40 astronomical camps have been organized by the society so far. More than 20 have been dedicated to observing meteor showers.

In May 1997, International Meteor Organization (IMO) President Jirgen Rendtel and two other IMO members visited the JAS during the Eta Aquarius shower. The IMO delegation gave valuable lectures about observing techniques, including how to monitor meteors with an FM radio receiver and Yagi antenna, which can detect smaller meteors than visual observations, (http://proquest.uni.com/pqdweb). The most important activity related to meteor showers was the Leonid Conference organized by JAS in cooperation with IAASS, held from 12 to 21 November 1999. More than 40 distinguished astronomers attended the conference, among them Professors Jack Backly and David Fisher. During the first two days, more than 15 papers were presented at Al al-Bayt University on the subject of meteor showers. The participants then moved to Al-Azraq camp, which is located in a desert region 60 km east of Amman, at which they observed more than 4,000 meteors.

INSTITUTE OF ASTRONOMY AND SPACE SCIENCES (IAASS) AT AL AL-BAYT UNIVERSITY (AABU)

The establishment of IAASS (http://www.aabu.edu.jo/space/mainl.htm) at AABU (http://www.aabu.edu.jo) in 1994, came as a concrete step towards linking the University with developed countries worldwide and keeping it on track with scientific innovations and technological know-how in these countries. IAASS represents the nucleus for basic space science in the Arab world. It is hoped that it will continue to develop with a view to fulfilling the needs and aspirations of the University community, the local community and the region as a whole.

Objectives

- Preparing and qualifying national human resources to work in the two fields of astronomy and space sciences and promoting technological developments. The objective is to render services to Arab and Islamic societies in general, and to Jordanian society in particular

- Carrying out research, astronomical studies and observations in various wavelengths on various celestial objects

- Conducting research and studies of the atmosphere and its layers that may contribute towards improving the efficiency of wireless networks and management of the Earth's natural resources using satellite imagery

- Studying the ionosphere and magnetosphere and preparing mathematical models that can describe the motion of space plasma in those regions

- Sustaining the quality of graduate studies at the Institute with a view to supplying the local and Arab societies with human resources capable of assuming a number of specialized professional responsibilities in the future

Ways and Means of Fulfilling the Institute's Objectives

- Establishing scientific and technological links with foundations and centers in the Arab and Islamic world to secure maximum utilization and exchange of results, studies, and research carried out at the Institute

- Organizing conferences, seminars, scientific meetings, workshops and local training courses in cooperation with Arab, Islamic and international universities and scientific establishments, forming a specialized team that would help to promote basic understanding of space sciences and increasing participation in similar activities inside and outside Jordan

- Strengthening cultural and scientific ties with counterpart universities and institutions on issues of common interest. This may include signing scientific agreements, exchanging experiences, exchanging experts to follow-up on scientific and technological developments, exchanging updated information and offering consultations and scientific experience in relevant specialties to universities and scientific foundations.

•   Collecting information, data sources, and various documents that are necessary for the follow-up to relevant technological developments in areas of specialization

•   Securing the availability of specialists and assistants in the fields of astronomy and space sciences and keeping their knowledge and experience up-to-date

•   Installing and acquiring laboratories and workshops to enhance research activities at the Institute

•   Attracting human resources in order to create of a selection of scientists, researchers, staff members and technicians capable of working with astronomy and space sciences and related technologies

•   Strengthening the relationship between theoretical, applied and basic research projects conducted by Al al-Bayt University and other universities, on the one hand, and applied research conducted by research and development (R&D) departments in the public sector, on the other, with the objective of applying research projects to activities in the public sector.

The Institute of Astronomy and Space Sciences (IAASS)

The Institute of Astronomy and Space Sciences consists of a Department of Astronomy, a Department of Space Sciences, a Department of Environment and Water Resources (Strategic Environment and Water Resources Research Unit) and an Information and Computer Unit.

Department of Astronomy

This Department consists of the Department, with its teaching, technical, and administrative staff, the Astronomical Observatory and the Unit for Reviving the Astronomical Heritage of Arabs and Muslims.

The Department is entrusted with task of teaching astronomy for graduate students and carrying out theoretical and observational research and studies on different celestial objects and clusters using various wavelengths using varying observational and analytic facilities available in the observatory. It is anticipated that the Institute will cooperate with international observatories and organizations in these activities. In addition, the Institute carries out practical studies on designing observational devices and accessories, analytic and computer control and its future developments, and applications used in fixing the beginning of lunar months and prayers time.

Department of Space Sciences

The Department of Space Sciences works with space physics, remote sensing and space communications.

The Department is in charge of teaching at the graduate level and carries out research on the atmosphere. Its fields of research include the ionic layers in relation to wave propagation inside the atmosphere, which aims at utilizing commercial links and radio frequencies.

In addition to using available space technologies for studying the Earth's surface and its environment, the institute is involved in the study of aspects of space physics, such as cosmic rays, airglow, geomagnetism, ionosphere physics, the upper atmosphere, and geophysics. The Institute is also involved in the study of natural resources and their applications in Jordan and in building the basic infrastructure for image analysis and its applications in different fields.

Strategic Environment & Water Resources Research Unit

This unit (http://www.aabu.edu.jo/water/home.htm) is in charge of teaching at the graduate and post graduate level and for carrying out environmental research and surveys, including non-conventional research projects related to development of resources.

Information and Computer Unit

This Unit is responsible for building a computer network with monitors connected to other units and departments at the Institute for the analysis of scientific data and satellite images. The Unit will also store information concerning astronomical and space research projects, while keeping open communication lines with international institutions. This process is vital for getting acquainted with international research and studies in areas of specialization.

Curriculum

The study plan at the Institute is divided into nine credit hours for compulsory courses and nine credit hours for a thesis. In addition, the Institute offers six credit hours as university requirements and three credit hours as remedial requirements. The total number of credit hours needed to obtain an M.Sc. degree is 33.

The contents of compulsory and elective courses in the various fields of astronomy and space sciences are in line with those offered by similar departments in international universities. The compulsory courses concentrate on the fundamentals of astronomy and space sciences, mathematical physics and astronomical and space techniques. Elective courses for astronomy students, on the other hand, concentrate on astrophysics, radio astronomy, cosmology, celestial mechanics, stellar structure and galactic and extragalactic structure, while elective courses for the space sciences students include space physics, remote sensing and related technologies, plasma physics, electromagnetic theory, computational techniques and radio astronomy.

To be eligible for admission to the M.Sc. Program offered by the Institute, candidates should be holders of a B.Sc. degree in one of the following fields: astronomy; physics; mathematics; space engineering; electrical engineering or communications and control system engineering.

Academic Facilities (Equipment and Observatory)

The Institute's laboratories, equipment and academic facilities include the following:

• The Computational Laboratory – This consists of a network equipped with terminals that provide the academic departments and units at the Institute with data significant to research in space and astronomy and with studies relevant to the concerns of the Institute.

• Remote Sensing Laboratory – This includes a number of devices, in particular a number related to computer scanning and to remote sensing images and their analysis. It uses software such as Remote Sensing ERDAS Software and DRISI Geographic Information System Software.

Astronomical Observatory (Maragha Observatory)

The Meade 16" LX200 Telescope

The Meade Schmidt-Cassegrain Optical System: In the Schmidt-Cassegrain design of the Meade 16" model, light enters from the right, passes through a thin lens with two-sided aspheric correction ("correcting plate"), proceeds to a spherical primary mirror, and then to a convex aspheric secondary mirror. The convex secondary mirror multiplies the effective focal length of the primary mirror and results in a focus at the focal plane, with light passing through a central perforation in the primary mirror.

The 16" mode includes an oversize 16.3759" (about 41.59 cm) primary mirror, yielding a fully illuminated field of view significantly wider than is possible with standard-size primary mirrors. It is this phenomenon which results in Meade 16" Schmidt-Cassegrains having off-axis field illumination 10% greater, aperture for aperture, than other Schmidt-Cassegrains utilizing standard-size primary mirrors.

Main Specifications and Features, 16" LX200:

Optical Design	Schmidt-Cassegrain
Clear Aperture	406.4 mm (16")
Primary Mirror Diameter	415.9 mm (16.375")
Focal Length	4064 mm

Resolving Power (arc sec.)	0.28
Limited Photographic Magnitude	18.0
Image Scale (arc sec/mm)	50.75
Maximum Practical Visual Power	800
Telescope Mounting	One-piece fork, double-tine
Materials: Tube body	Aluminum
Mirrors	Pyrex glass, Grade-A
Correcting Plate	BK7 optical glass
Telescope Dimensions, swung down	18" x 26" x 51"

CCD Camera (Pictor 1616)

The observatory contains, in addition to the telescope, a Charge Coupled Device (CCD), which can be used for photographic, photoelectric and spectral observations of celestial objects. The main specifications are:

CCD size	1536 x 1024
Pixel size	9 microns
Raw image size	3 MB
Dark current (20 deg. C)	< 1 e$^-$/5 seconds
Quantum efficiency	$< 30\%$ at 520-700 nm.
Readout noise	< 15 e$^-$ rms.
Charge transfer efficiency	99.999%
Well depth	85,000 e$^-$ per pixel
Dynamic range	65,536 brightness levels
Readout time	10 microns per pixel

Host PC programs include image transfer and display, contrast control, smoothing, sharpening filters, and image alignment.

On board programs include automatic exposure level, dark current exposure and flat-field sequence, pixel binning and acceptance of a color filter system.

Computer

The telescope and CCD are controlled by a Pentium II 440 BX computer, which can also be used for data acquisition. The astronomy software used is the sky level IV and epoch 2000 image processing with interface cable.

Main Purposes of the Astronomical Observatory

- Training of AABU students, as well as students from other universities in Jordan
- Research projects and observations for M.Sc. and Ph.D. students
- Research projects and observations for the Institute staff
- Unit of the network of small astronomical telescopes of the World
- Training of school students and development of astronomical education in Jordan

- The observatory will be used for observing celestial objects, mainly variable stars (Eclipsing Variables) and some astronomical events, including the crescent observations
- To promote scientific cooperation, any Arab or international astronomer may use the observatory (Al-Naimiy & Kandalyan 2000a).

The Observatory site

The observatory is located at the campus of AABU, near the building of the Institute of Astronomy and Space Sciences.

Longitude: 36° 14' 19"
Latitude: 32° 20' 30"

The field of view at the observatory is 2-3 arcsec. The number of clear nights is around 200 days per year.

The name of the Observatory: Maragha

The name "Maragha" comes from the name of an old observatory, which was installed by the Muslim astronomer Nasir Al-Dean Al-Tousi in 627 H/1264 A.H. in the city of the Maragha, the capital of the old Azerbaijan. The old observatory contained mostly books that had been brought from Iraq, Syria and Al-Jezera during Baghdad's collapse in 656 H/1258 A.H. Many Islamic astronomers used the observatory during that time.

Aspirations of the Institute

The Institute aspires to realize the following projects and attain the following goals and objectives in the future:

Institute Departments

In the next five-year plan, the Institute hopes to establish the following divisions and departments:

- Astronomy Department and Al al-Bayt Observatory
- Space Physics Department
- Remote Sensing and Environment Department
- Communication and Meteorological Department

It is anticipated that the following laboratories, workshops and telescopes will support these departments:

- Mechanical and Electronic Workshop
- Remote Sensing Laboratory
- Communication Laboratory

- Meteorological Laboratory
- Computer Laboratory
- Optical and Radio Telescopes (establishing a 1.5m optical telescope and 20m radio telescope in a selected site in Jordan, which gives Jordan a good opportunity for fruitful cooperation with other astronomical institutions from different countries)

International Cooperation

The present and future plans of the Institute are oriented towards opening wide channels of communication with other parts of the world. This may be fulfilled by means of scientific agreements, personal contacts and participating in conferences and related scientific activities. The Institute also aspires to enhance its cooperation with related scientific international organizations such as the International Astronomical Union (IAU) and International Geophysical Union (IGU). Accordingly, IAASS maintains very good scientific relations with the United Nations (UN) and the European Space Agency (ESA) through the series of UN/ESA workshops on Basic Space Science.

Teaching, Scientific Research and Technical Staff

Given the fact that research techniques and technologies in space sciences are developing at a high pace, the Institute has plans to achieve the following:

- Select distinguished scientists of a high standard that are up-to-date with developments in the field of AASS at the international level

- Fully utilize all available avenues of training such as scholarships, sabbatical programs, and training courses that may contribute to securing a sufficient number of staff of all needed levels and specializations

- Offer intensive training programs for technicians who would work on existing, specific projects on a contractual basis. The programs should include design and scientific activities for different technological aspects of these projects. This is needed in order to improve the skills of the technicians and to help them keep up with technological developments, as well as to satisfy research and teaching requirements.

Plans for the coming five years

The Institute's plans for the coming five years include the following:

- Employing about 20 researchers and teaching staff, who hold Ph.D. degrees in the fields of space physics, astrophysics, astronomy, cosmic ray physics, atmospheric physics, electronic engineering, communications engineering, space image analysis, remote sensing and computer science

- Recruiting about 30 staff assistants such as engineers and physicists, who would hold B.Sc. or M.Sc. degrees, in addition to a number of technicians to organize and operate the observatory, organize workshops, and run intensive and integrated training programs in collaboration with international organizations.

Selection of projects

In order to maintain the quality of teaching and research, the Institute plans to move in two directions:

• Ready-made Projects

This includes signing contracts with international agencies for designing, manufacturing and establishing large and developed scientific research facilities that can serve two goals: 1) Providing staff members with state-of-the-art technologies needed for conducting excellent research programs in accordance with the latest advances worldwide, and 2) Proving to the international community the academic seriousness of AABU in pursuing scientific research and seeking knowledge, and encouraging international organizations, committees and renowned scientists to cooperate with the Institute.

• Local Projects

Local projects would be initiated, designed and carried out completely or partially by staff members of the Institute. Given the limited experience and facilities available at the Institute, projects at the very beginning will be modest. They will, hopefully, develop and expand in the future, as the Institute is increasingly able to depend on its own resources.

Future Scientific Projects

1.   Establishing new mathematical models relevant to space physics, the ionosphere, the magnetosphere and the upper atmospheric layers.

2.   Establishing a catalogue of observed x-ray stars. This would include physical and geometrical properties of these stars and their positions in the sky using new mathematical methods and computer programs of an international standard.

3.   Establishing an astronomical map for the Hashemite Kingdom of Jordan including correction of the al-Qebla direction for the Kingdom's mosques, tables of prayer times and calculations of the beginning of the lunar months for the next hundred years, especially the religious months such as Ramadan, Shawwal and Thu al-Hujja.

4.   Studying different aspects of natural resources for the Arab and Islamic world using satellite image analysis and distributing the results to the concerned countries according to their interest.

5.   Building a large optical and radio Astronomical Observatory (AO), with an optical mirror at least 1.5 meters in diameter and a radio dish at least 20-30 meters in diameter.

6.   Keeping the Institute up-to-date with international scientific projects and taking part in some of them, especially those relating to studies of stars with a complete spectrum. This can be easily accomplished now that the Institute has obtained a small optical telescope (40cm). In addition, the Institute is now hoping to enhance its cooperation in this field with Arab countries.

7.   Further plans will be made to offer an academic program leading to a Ph.D. degree in AASS.

8.   IAASS aims to expand its activities at the local and regional levels. This can be done by improving links with the Jordanian public and regional universities and by investing in the corresponding experimental capabilities that are already available at those universities.

IAASS Activities

The most significant activities of the Institute in the last three years included holding a number of workshops in Space Communications, Remote Sensing (Science and Technology), Astrophysics and Space Physics. In addition, the Institute organized many astronomical observation nights using the Maragha Observatory, one of which was devoted to observing the Hale-Bopp comet and another of which was devoted to observations of planets and a few variable stars. These events were attended by a large number of distinguished physics professors, engineers and other faculty from Jordanian universities and scientific centers. The Institute organizes a continuing seminar series which has been attended by interested staff members and graduate students.

The most important activities include the following:

•   The First International Conference was held at the IAASS of the AABU in Mafraq, Jordan from 4 to 6 May 1998. The conference was attended by more than 50 astronomers and space scientists from 16 countries and entities: Algeria, Australia, Belgium, France, Germany (UN), Iran, Iraq, Jordan, Kuwait, Mexico, Palestine, Sudan, Tunisia, Turkey, the United Kingdom and the United States of America. The program of this Conference included presentations on: The Earth's Environment; The Solar System; Sun and Star Formation and Binary Stars; and Cosmology and Missions to the Universe. The proceedings for the conference have been published by Al al-Bayt University press (Al-Naimiy & Kandalyan 2000a).
        (http://www.seas.columbia.edu/~ah297/un-esa/ws1998-jordan-astronomy.html)

•   IAASS Hosted the Eighth United Nations/European Space Agency Workshop on Basic Space Science, which was co-sponsored by the United Nations, the European Space Agency and the Government of Jordan and hosted by the Al al-Bayt University in Mafraq, Jordan, from 13 to 17 March, 1999. More than 120 astronomers, scientists and students of basic space sciences from 35 countries attended the workshop. The proceedings for the workshop will be published as a volume of the journal "Astrophysics and Space Science".

(http://www.seas.columbia.edu/~ah297/un-esa/activities.html#1999)

ARAB UNION FOR ASTRONOMY AND SPACE SCIENCES (AUASS)

(http://www.jas.org.jo/union.html) or (http://www.auass.org/srd)

This Union was established in August 1998 as an outcome of the second Arab Conference on AASS, held in Amman, Jordan from 29 August to 1 September 1998. The Conference was organized by JAS and IAASS. The participants included more than 100 astronomers and scientists from 14 Arab countries, in addition to observers from France, Italy and the United States of America. The participants decided to establish the Union with headquarters in Amman. The aim of the Union is to develop AASS in Arab countries, for instance through conferences, meetings, publications and joint research projects, in cooperation with international AASS institutions.

Objectives

- Introducing astronomy and space science in the Arab countries as scientific disciplines playing a role in the scientific and technical progress and development of the Arab world

- Encouraging the exchange of scientific data, information, and expertise in various fundamental and applied fields of astronomy and space science among concerned institutions and individuals

- Improving the scientific and technical competence of specialists in astronomy and space science working in Arab countries

- Standardizing scientific and technical terms in astronomy and space science that are in use in the Arabic language

- Researching the contribution of the Arab Islamic civilization in the realm of astronomy and related sciences, as well as the roles of various other ancient civilizations in the region

Ways and Means of Fulfilling the Objectives

- Promoting astronomy education as a pivotal means to fulfill the long-term objectives

- Organizing conferences, seminars, and workshops in astronomy and space sciences

- Encouraging the popularization of astronomy and space sciences through writing, translating, and publishing articles and books for the general public, through public lectures and through other means

- Creating databases containing information about professional institutions and societies dealing with astronomy and space science in Arab countries, about Arab specialists in these fields and also about astronomical societies, clubs and planetariums in the Arab World

- Sponsoring some long-term (large-scale) astronomical projects that involve the participation of as many Arab specialists, institutions, and countries, as possible

- Developing cooperation at the international level, with the aim of realizing AUASS's objectives

Some Activities of the AUASS

The AUASS has established a few scientific committees such as:

- The Crescents Observation and Mawaqeet Committee of the AUASS, to consider questions relating to the young crescent moon visibility prediction (for the use in fixing the beginning of the Islamic Holy Lunar months). The Committee has created a refereed Journal called the Arab Journal for Astronomy and Space Sciences (AJAASS), to be issued twice a year starting from September 2000.
(http://www.jas.org.jo/icop.html)

- The scientific research division (SRD). The SDR aims to help the AUASS achieve its general goals and objectives, particularly those connected to scientific research in the various fields of AASS. This can be achieved principally through the creation of various working groups within the SRD, by enhancing the scientific collaboration between the SRD's members via the exchange of astronomical data and information, and by finding research projects of common interest that might eventually be adopted by the AUASS's Higher Council and thus supported by the AUASS's member states. Although the SRD/AUASS is primarily concerned with research in AASS at the basic and applied levels, it should be noted, in this era of multidisciplinary research, that the interests of the SRD/AUASS cannot (and should not) be limited to basic science. In other words, the emergence of working groups in somewhat social fields, such as the history of AASS, or even AASS in culture and education, might seem inevitable at some stage. In fact, any working group created within the SRD/AUASS would reflect the research interests of its members, as long those interests fit with the AUASS's goals and internal regulations.

ISLAMIC CRESCENTS OBSERVATION PROJECT (ICOP)

The ICOP is a global project, organized by the AUASS and JAS. It aims to gather the largest possible number of lunar observers worldwide. The project's purpose is predicting the visibility of the young crescent moon which fixes the beginning of the Islamic Holy Lunar months (http://www.jas.org.jo/icop.html). ICOP is supervised by a committee from AUASS, IAASS and JAS. Members come from a number of different Islamic countries (http://www.jas.org.jo/come.html).

Objectives

• To develop the crescent's observation criteria, based on the data and information obtained by participants in this project, to improve judgment of the visibility of young crescents

• To inform people around the world of the first day of some important lunar (Islamic) months, particularly the Holy month of Ramadan, and Shawwal, which is of particular interest for those who do not live in Islamic countries

TEACHING OF ASTRONOMY AND SPACE SCIENCES IN JORDANIAN SCHOOLS AND UNIVERSITIES

Most secondary schools have a general astronomy course. The courses contain general information about the Earth-Moon system, day and night, seasons, the Solar System, stars, clusters, the Milky Way, galaxies and finally the Universe. In addition, astronomy and astrophysical courses at the B.Sc. level are offered by the physics departments of most Jordanian universities.

FUTURE EXPECTATIONS

An important decision has been made by United Nations, in that the Hashemite Kingdom of Jordan has been selected to host the regional Centre for Space Science and Technology Education for Western Asia. This will be an excellent move toward developing AASS in the Arab countries.

Jordanian astronomical organizations have a strong desire to move towards building a large optical and radio astronomical observatory, with an optical mirror at least 1.5m in diameter and a radio dish at least 20m in diameter, to be used by local and international astronomers and scientists.

In cases where large astronomical and space science research projects are too expensive for a single country to support on its own, groups of countries may establish a common space agency, an example of which is the European Space Agency (ESA). Therefore, we hope that the Arab and Islamic countries can join hands to establish an Arab Space Agency (ASA).

REFERENCES

Al-Naimiy, H. M. K. (2000). "The Importance & Needs of AASS in Arab Countries", in press.

Al-Naimiy H. M. K. and Kandalyan, R. (2000a). "Maragha Astronomical Observatory", in press.

Al-Naimiy, H. M. K. and Kandalyan, R. (2000b). "Proceedings of the 1st International Conference in AASS, 4-6 May, 1999".

Ohda, M. S. "Sky & Telescope", December 1996.

ASTRONOMICAL SITE TESTING WITH A METEOROLOGICAL SATELLITE – THE FIRST STEP FOR THE IMPLEMENTATION OF A NETWORK OF ROBOTIC TELESCOPES[*]

Ahmad Tamer Al-Mousli
General Organization of Remote Sensing (GORS)
P. O. Box: 12586
Damascus, Syria
Phone: 00963112218765/00963112218764
Fax: 00963113910700
E-mail: gors@mail.sy

A   ABSTRACT

Plenty of satellites are orbiting around our home engaged in surveying, observing and inventorying the Blue Planet. They take a close look at our forests, water resources, crops, vegetation, fauna and flora, soil types, urban and rural situations, record and assess natural disasters, observe the environment and document the hole in the Ozone Layer.

All these satellites send their data to users via robotic ground stations and the data can then be used to provide information to decision makers.

B   INTRODUCTION

Remote sensing means the science and art of obtaining information about an object or phenomenon through the analysis of data acquired by a device that is not in contact with the object, area or phenomenon under investigation.

Satellite remote sensing is the collection of information about the Earth using sensors that are installed on spacecraft, information which is sent to ground stations. After this, the information is corrected and enhancements are made, including geometric corrections, image enhancement, atmospheric corrections, contrast enhancement, density slicing and filtering.

---

[*] *This paper was presented at the "Ninth United Nations/European Space Agency Workshop on Basic Space Science", held from 27 to 30 June 2000 in Toulouse, France, and does not necessarily reflect the views of the United Nations.*

After this process, we obtain an analog image from digital image. This analog image is used in the following stages of analysis:
- Unsupervised classification.
- Supervised classification.

Unsupervised classification is carried out before before ground truthing and can be determined by the tones, texture and isolation of the polygons. Supervised classification, carried out after we have isolated the polygons, is the process of going into the field to check all types of polygons. Each kind of polygon means something different.

C   APPLICATIONS OF REMOTE SENSING

The applications of satellite data listed below were carried out by special institutions in Syria, including the General Organization of Remote Sensing (GORS).

1   Cartography:

- Production of maps on the scale 1:100,000 and smaller
- Topographic maps
- Thematic maps for various fields of application
- Plotting of contour lines from stereoscopic satellite data for the scales 1:100,000 and 1:50,000
- Map revision
- Digital cartography
- Establishment of land information systems
- Preparation of a Digital Terrain Model (DTM)
- Merging topographic maps and aerial photos for a project to connect the first interchange with the fifth interchange of the southern highway

2   Agriculture:

- Agriculture resources inventory
- Preparation of agricultural statistics
- Soil mapping
- Monitoring of soil erosion
- Seasonal vegetation changes
- Fruit freezing
- Cultivated regions
- Irrigated and drained agricultural land
- Detection of overgrazed and saline areas
- Crop forecasting
- Agriculture sector assessment in Daraa province

3   Forestry:

- Forest mapping
- Forest inventory
- Monitoring of:
  - changes of forest stands
  - erosion sensitive areas
- Detection and monitoring of forest

4   Climatology:

- Weather forecasting
- Detection of air pollution
- Disaster forecasting
- Preparing weather maps for traffic (aviation, shipping)
- Agro-climatic monitoring

5   Geology:

- Compilation of geological, pedological and hydrogeological maps
- Analysis of lithology and geological structure (lineament, folded structures)
- Determination of potential ground and surface water
- Suitability of soil for land use
- Evaluation of data for development of infrastructure
- Study of iron and other mineral deposits using space imagery

6   Regional planning:

- Compilation of maps relating to:
  - Infrastructure
  - Population density
  - Structure of settlements
- Determination of land use
- Determination of land occupation in urban areas
- Climatic impact of land use changes
- Urban planning of Aleppo and the surrounding region

7   Environmental Monitoring:

- Disaster forecasting
- Determination of damage after a disaster
- Monitoring of various types of pollution
- Monitoring of desertification
- Observation of volcanic eruptions
- Evaluation of the environmental impacts on the Damascus Basin

8   Water Resources:

- Monitoring river flooding
- Build-up and subsequent abatement
- Monitoring pollution
- Determination of water basin boundaries for rivers, used to mitigate flooding
- Determination of the water budget
- Mapping of coastal zones
- A snow cover study on the Al-Hermon mountains

D   WHY DO WE USE REMOTE SENSING IMAGES?

- Direct to use
- Inclusiveness
- Repetition
- Cheaper
- Accuracy

E   THE CONDITIONS OF THE ASTRONOMICAL SITES

- Elevation
- Transparency
- Road network
- Interference (electromagnetic pollution and light pollution)

F   METHODOLOGY

Two satellite images on the scale 1:50,000 were used from the Landsat Thematic Mapper (TM), which were acquired in summer 1995 and prepared in a photolab (GORS) in false composite color (FCC), with three spectral channels:

Ch 2	520-600 nm
Ch 3	630-690 nm
Ch 5	1550-1750 nm

We also used topographic maps, geological maps and a Climatic Atlas of Syria.

The orbital parameters of Landsat TM are as follows:

Orbit	Near-polar, Sun-synchronous
Altitude	705 km
Inclination	98.22
Repeat cycle	16 days

Applications of channels on Landsat TM are as follows:

Channel (Ch)	Wavelength (nm)	APPLICATION	Resolution (m)
Ch 1	450-520	Water penetration; Discrimination between soil and vegetation and between deciduous and coniferous forest	30
Ch 2	520-600	Measurement of visible green reflectance peaks of vegetation, used for vigor assessment	30
Ch 3	630-690	Chlorophyll absorption i.e. Vegetation discrimination	30
Ch 4	760-900	Biomass content; Delineation of land and water	30
Ch 5	1550-1750	Moisture content, Discrimination between snow and clouds	30
Ch 6	10400-12500	Thermal; Vegetation stress; Moisture determination	120
Ch 7	2080-2350	Discrimination between rock types; Hydrothermal mapping	30

G   RESULTS

The following table illustrates the work carried out:

Site	E	N	Al	Precip (mm)	T max	T min	Wind	Hum %	Geo	Farms	LP	Road
1	36 28 38.8 8	34 01 47.4 1	2616	250-300	15-16	4-6	SW N-W S-w S	55-60	Lime-stone	poor	-	+
2	36 05 22.6 8	33 36 27.3 4	1319	300-350	19-20	6-8	W NW E-W E	50-55	Lime-stone	plenty	-	+
3	36 22 13.4 8	33 43 40.2 5	1910	250-300	16-18	8-9	W NW E-W E	45-50	Lime-stone	poor	-	+
4	36 26 53.1 4	32 52 42.1 6	771	350-400	22-23	8-10	W W S-W S	55-60	Bas-alt	poor	-	+

H   SUMMARY

We can summarize the paper in these four points:

<u>Location 1</u> – Near Essal Alwared

This location is the most suitable, as determined by analyzing satellite images and topographic maps.
- There is a low density of towns.
- There is only sparse vegetation cover.
- Because of the elevation, 2616m above sea level, and the length of the western Lebanon chain, which contains some tops above 3000m above sea level, the conditions are better for haze.
- There is a new road.

<u>Location 2</u> – Near Al Nabi Habil.

This location is situated near to a high density of settlements and towns. The elevation is 1319m above sea level but there are many higher peaks. There are plenty of farms and there is dense vegetation cover near this location because of the nearby Barada River.

<u>Location 3</u> – North of Sednaia – Accepted.

<u>Location 4</u> – Near Najrran.

There are plenty of farms and lakes situated to the west of this location.

REFERENCES

1.   Remote Sensing for Development. Proceedings of the International Conference in Berlin (West) September 1986.

2.   Space-Based Remote Sensing. A Report to the Congress, September 1987.

3.   Climatic Atlas of Syria.

4.   Geological Map of Syria.

A PLAN FOR BUILDING AN INTERNET-ACCESSIBLE ROBOTIC OBSERVATORY IN MALAYSIA[*]

Mohammad Ridwan Hidayat
Space Science Studies Division, MOSTE
Planetarium Negara, 53 Jalan Perdana
50480 KUALA LUMPUR
E-mail: ridwan@baksa.gov.my

ABSTRACT

In conjunction with the Internet Based Educational Resources (IBER) project, the Space Science Studies Division (BAKSA) of the Malaysian Ministry for Science, Technology and the Environment is planning to build a web-enabled robotic observatory that would operate a remotely controlled telescope and charge coupled device (CCD) camera in a real-time and provide a hands-on and interactive environment to students and research scientists, especially in Malaysia. Scripting software would allow the telescope and CCD camera to do the task automatically. This will enrich the space science education program as well as professional astronomical research activities in Malaysia. This paper will look at conceptual design for several existing robotic systems available on the Internet and will present the proposed system, which is suitable for conditions in Malaysia.

1. INTRODUCTION

The availability now of powerful CCD cameras and automated telescopes has revolutionized observational astronomy, because of the technology's precise, simple, economic and efficient operation. The observer can remotely control the telescope and CCD camera from a convenient place, either in real-time or in robotic mode. This development is supported by the availability of the Internet, which is connected globally. A 'merger' of available automated telescopes around the world could be created via the Internet, which would have a lot of advantages. More detailed information on the advantages of robotic telescopes can be found in Othman and Hidayat (1995).

In conjunction with the IBER project, BAKSA has decided to build a web-enabled robotic telescope and CCD imaging system to provide practical observation opportunities for students and research scientists to study the universe. This project will be embarked upon in two stages:

- The first is to automate the existing telescope system in Kuala Lumpur by equipping it with a robotic mount. In this stage, we will gain experience in managing and

---

[*] *This paper was presented at the "Ninth United Nations/European Space Agency Workshop on Basic Space Science", held from 27 to 30 June 2000 in Toulouse, France, and does not necessarily reflect the views of the United Nations.*

controlling the telescope system with the complexity of a high number observing applications.

- The second stage is to build a professional grade robotic observatory for more in-depth research projects. This observatory will be located at the best site available in Malaysia and will be fully robotic. This means that the telescope will not be supervised either locally or remotely during its routine operation (Steele 1999).

2. AVAILABLE AUTOMATED/ROBOTIC TELESCOPES

It is interesting to note some of the advantages of remote robotic operation of telescopes, which include (Liverpool Telescope 2000):

- Running costs reduced – no staff on-site;
- Lower capital cost;
- Observations carried out efficiently and in a predictable and repeatable way;
- Many different programs can be undertaken each night with observations optimized to observing conditions;
- Effective exploration of the temporal domain (targets of opportunity, long-term monitoring, multi-frequency campaigns);
- Ideal for remote observing as part of an educational program.

There are two types of telescope control system that utilize an Internet connection: real-time telescope controlling; and a fully robotic system, pre-programmed with dynamic priority scheduling.

2.1 Real-time remote control using scripting software

In this system, the user can control a remote system in real-time and the result can be downloaded quickly after the observing session has ended. Available automated/robotic telescopes with complete documentation are listed below:

Telescopes In Education (TIE), Mount Wilson Institute
http://tie.jpl.nasa.gov/tie/

Berkeley Automated Imaging Telescopes (BAIT), University of California at Berkeley
http://astron.berkeley.edu/~bait/

2.2 Fully robotic, pre-programmed with dynamic priority scheduling

The second type of telescope control system uses Internet-base forms, which many users can afford since there is no need to buy specific software. Their operation depends on priority scheduling and the result can be downloaded the day after the observing session. Available automated/robotic telescopes with complete documentation are listed below:

Bradford Robotic Telescope, University of Bradford
http://www.eia.brad.ac.uk/rti/index.html

The Liverpool Telescope of John Moore University (JMU), UK
http://telescope.livjm.ac.uk/

The Iowa Robotic Observatory (IRO), University of Iowa
http://denali.physics.uiowa.edu/IRO/

3. PROJECT IMPLEMENTATION

BAKSA will embark on the automated/robotic observatory project in two stages: preliminary and grand planning. All the systems must have three-control modes: manual; semi-automatic; and robotic.

3.1 Preliminary Implementation – Paramount GT-1100

In the preliminary implementation, the existing telescope Celestron C-14 will be equipped with a Paramount GT-1100 robotic mount from Software Bisque. The main objectives of this stage are to observe certain celestial objects under the city lights and to gain experience managing and controlling the telescope system, especially managing the complexities of high numbers of observing applications. This instrument is located at the BAKSA Observatory in Kuala Lumpur.

The Paramount GT-1100 robotic mount and the Celestron C-14 are equipped with a SBIG ST-7 (non ABG) CCD camera and automated filter wheel. The Sky and CCDSoft astronomical software are used for telescope and CCD camera control. RASServer allows remote clients running RASClient to access the telescope and CCD camera via the Internet. The user must have RASClient, TheSky and CCDSoft in order to control the telescope and CCD camera remotely. Scripting software is used to control the telescope and CCD camera for automatic observing tasks.

As for the Mount Wilson Observatory's Telescope In Education (TIE) project, the user must have a minimum of the following equipment and software for remote operation (Bartosh & Duff 1997):
- IBM PC-type computer, 386 or better, running Windows or Windows '95;
- SVGA monitor and driver card – these will provide the best viewing of the high quality images that will be downloaded during the observing session;
- At least 4 megabytes of RAM and 8 megabytes for Pentium processors operating Windows '95 – this is necessary for operation of the software that controls the telescope.
- A 9600-baud or modem or faster
- Modem and telephone lines – a voice line, in addition to the modem line, is required to facilitate communication with the volunteer telescope operator during sessions
- The Remote Astronomy Software (RAS) package, including TheSky and CCDSoft.

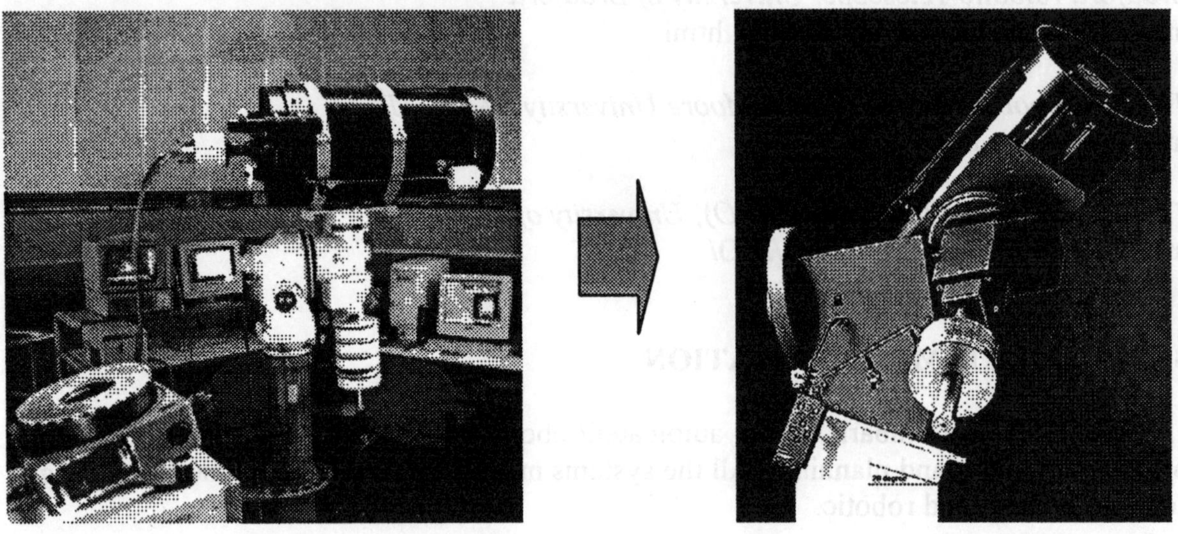

Figure 1. Inside BAKSA Observatory, Kuala Lumpur. Previous mounting system (left) and robotic mounting (right).

The next task is to make the dome rotation always synchronous with the telescope direction. Instant conversion between Ra-Dec to Alt-Az of the telescope movement can determine the direction of dome slit.

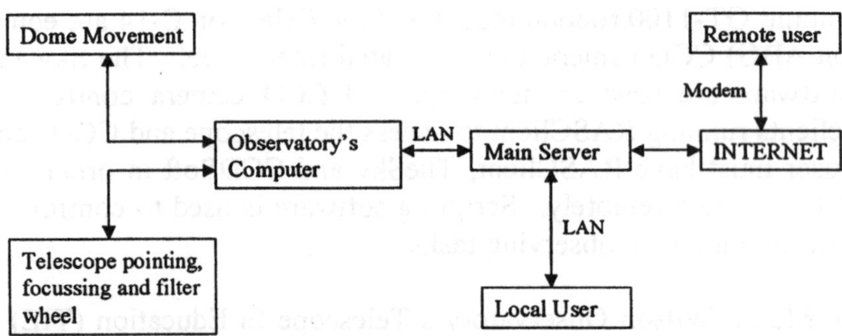

Figure 2. Block diagram of Paramount GT-1100 mounting, which will be implemented at BAKSA Observatory, Kuala Lumpur.

This is quite difficult to implement with the previous dome motor, because it uses an ordinary AC electric motor with the risk of slip when rotated in the dome. Applying a gear system between the motor and the dome could solve this problem. Further information about the robotic mount and RAS software is available on Software Bisque's web page (http://www.bisque.com).

3.2   Grand Planning: Professional Grade Robotic Observatory

A robotic telescope with a diameter of at least 0.5m is planned to be built in conjunction with the National Observatory facility. This will have a web-enabled control system and will be equipped with an Apogee AP-7 CCD camera. The remote user needs no specific software, only an Internet browser.

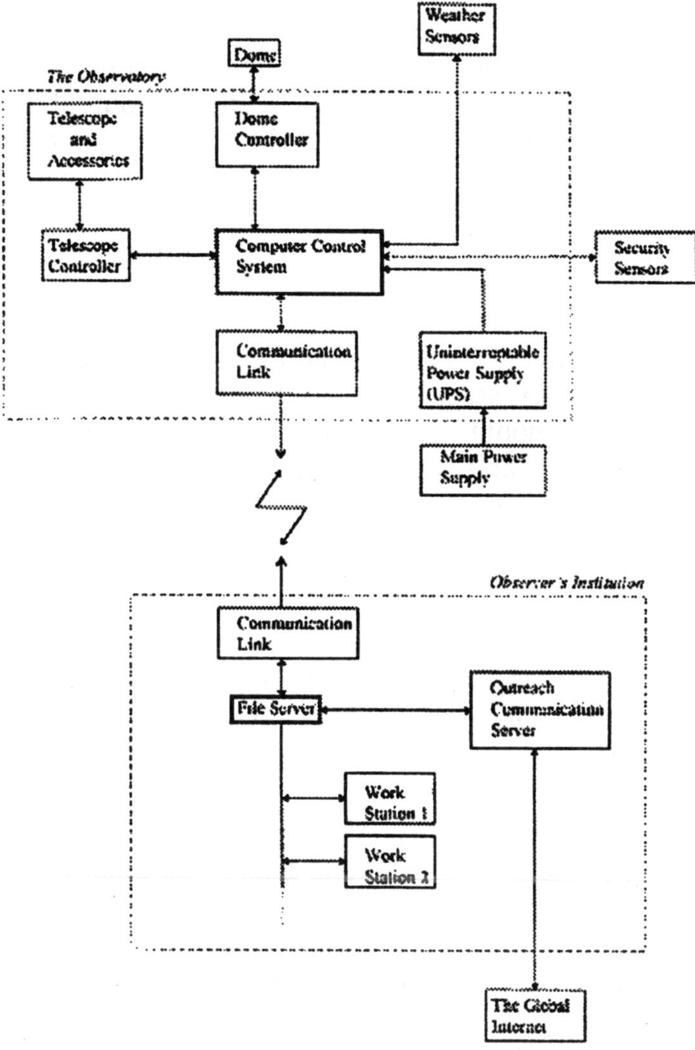

Figure 3. Block diagram of the proposed BAKSA robotic observatory, from Othman and Hidayat (1995).

The robotic observatory controls will provide the following tasks (Liverpool Telescope 2000):

- Scheduling observations
- Start-up and close-down procedures
- Weather monitoring
- Sequencing within a given observation
- Top level error handling and reporting

The control system of the Bradford Robotic Telescope, the Iowa Robotic Observatory and the JMU Liverpool Telescope will be studied to implement the proposal for BAKSA Robotic Observatory. The search for the best site for this Observatory in Malaysia is still ongoing.

4.   CONCLUSION

BAKSA is planning to build a robotic observatory, a project that will be embarked upon in two stages: a preliminary stage; and an advanced stage. In the first stage, a Paramount GT-1100 robotic mount will replace the old mount at the BAKSA Observatory in Kuala Lumpur and will act as a 'test-bed' before building the professional-grade robotic observatory.

REFERENCES

Bartosh, B. & Duff, B. 1997. *The Telescopes In Education (TIE) Users Guide and Workbook for Scholastic Programs and Amateur Studies, Ver. 2.1, Mount Wilson Institute. [http://tie.jpl.nasa.gov/tie/]*

Othman, M., Hidayat, M.R. 1995, *An Automated Telescope for Malaysia*, Proceedings of Third East-Asian Meeting on Astronomy: Ground Based Astronomy In Asia (ed. Norio Kaifu), National Astronomical Observatory of Japan.

Steele, I. A., 5 May 1999, *The Liverpool Robotic Telescope*. Elsevier Preprint.

The Liverpool Telescope of John Moore University, UK
[http://telescope.livjm.ac.uk/]

COMETARY AND SOLAR OBSERVATIONS WITH SMALL TELESCOPES CONNECTED TO A COMPUTER NETWORK[*]

Debi Prasad Choudhary
Udaipur Solar Observatory, Physical Research Laboratory
Post Box: 198, Udaipur – 313 001, India

INTRODUCTION

Optical telescopes are classified according to the size of their objective lens or mirror, depending on whether it is a refractor or reflector. The optical telescopes existing throughout the world at this time may be classified into four groups, namely "small" – up to 1 meter, "moderate" – between 1 and 3 meters in size, "large" – for the sizes between 3 and 6 meters, and "giant" – for telescopes larger then 6 meters. Small telescopes are playing a very important role in modern astronomical research with the advent of modern detectors and computer hardware and software (Sagar, 2000). Sagar (2000) deals mostly with astronomical observations related to rapidly varying astronomical sources such as γ-ray bursts (GRB's) and microlensing phenomena. Here, I shall describe the utility of small telescopes in observing large-scale cometary phenomena and solar activity. I shall also discuss the role of these telescopes and their cost effectiveness for astronomical education in developing countries.

The contribution of small telescopes to astronomical research can best be illustrated by the important and unique observations made by small telescopes situated in India. There are eight astronomical observatories spread over the range of latitudes in India. India is situated in the middle of a roughly 180°-wide longitudinal band without many modern astronomical instruments. The other astronomical telescopes are situated in eastern Australia in the east (~20° west) and in the Canary Islands (~157° east). As, a result, several unique and important observations, such as the ring of Uranus (Pandey et al., 1984) and GRB afterglow (Sagar et al., 1999), have been made from these observatories.

THE ADVANTAGE OF SMALL TELESCOPES

Obviously, small telescopes cannot replace large and giant telescopes. Nevertheless, this class of telescopes has an important role to play in view of the telescopes' cost effectiveness, efficiency, geographical location and other special requirements of astronomical observations. In this section, I shall briefly discuss these topics.

---

[*] *This paper was presented at the "Ninth United Nations/European Space Agency Workshop on Basic Space Science", held from 27 to 30 June 2000 in Toulouse, France, and does not necessarily reflect the views of the United Nations.*

Cost Effectiveness: Cost effectiveness can be defined as the large returns compared to the money spent on building the small telescopes and their running cost with moderate instrumentation such as photometers and spectrometers. It has been shown that large and giant telescopes are disproportionately expensive in this respect (Disney, 1972). It has also been shown that the number of research papers produced per year from the observations with small telescopes are higher compared to large or giant telescopes (Warner, 1986). If that reflects the annual return of the invested capital than it can be concluded that four half-size telescopes are as profitable as one full size telescope. Of course, caution must be used when comparing the nature of science that telescopes of two different classes produce. The cost of both these telescopes is same. We, at Udaipur Solar Observatory, are proposing a Multi Aperture Solar Telescope (MAST), precisely because of these considerations (Venkatakrishna, et al, 2000).

Efficiency: Being moderate- to low-cost, small telescopes can be equipped with dedicated instruments and used for specialized observations. In addition, when a back-end instrument is permanently installed on a dedicated telescope, it ensures stability of performance, maintains the calibration and saves time for instrument change.

Geographical location: The Indian longitudinal zone is situated in the middle of a roughly 180°-wide longitudinal band without many modern astronomical instruments but with clear sky. Similarly, the wide African longitudinal belt is devoid of many telescopes compared to its geographical extent. Installation of smaller telescopes at these longitudinal locations will serve greatly in detecting the time-varying and location-specific phenomena mentioned earlier in the case of GRB's.

Other special requirements: Some astronomical studies require observations over a long period of time. Sagar (2000) discusses this issue in the context of galactic astronomy. In the case of cometary astronomy, there is a great need for continuous observation when monitoring outburst events and solar wind phenomena. Similarly, in solar physics it is important to monitor photospheric and chromospheric events continuously in order to study their evolution. These aspects are described in the following sections.

COMPUTER NETWORKING

Today, computer networking of telescopes has become very easy and has consequently enhanced the power of these instruments. Computer networking can be carried out in two ways. (1) The first method is through communication between users of the individual telescopes located at very large distances from each other. Observation can be carried out using these telescopes in a coordinated fashion. For example, in 1989, when I was observing the sodium cloud of Io with the 2.2-meter telescope at Calar Alto, Spain, our collaborators in Italy were taking the spectra of the same region. We were communicating by telephone at that time. The same operation can be made faster with less cost using the Internet. An example of a group of telescopes connected over the Internet is the Global Oscillation Network Group (GONG) (http://helios.tuc.noao.edu/

homepage.html). This project consists of six telescopes situated at six locations in different countries, as shown in Figure 1, which observe the Sun continuously. The aim of the project is to study helioseismology in order to understand the internal structure of the Sun. (2) A combination of telescopes can be connected through a computer network for various sophisticated operations such as beam combining and simultaneous observations of several physical parameters. An example of such a project is described in MAST document (http://www.uso.etnet/mast).

COMETARY PHYSICS

Comets are made of "dirty ice" and contain the material of which the primordial Solar System was made. As they approach the Sun, the ice evaporates and interacts with the solar wind and radiation. The evaporated material produces two types of cometary tail, namely plasma and dust tails. The structure and dynamical signatures of the plasma tail depends on the local solar wind conditions. Therefore, the study of the cometary plasma structure helps in understanding the solar wind conditions of the interplanetary medium. Although interplanetary probes can be used to study interplanetary conditions, these are limited to near-Earth equatorial regions.

The plasma activity of comets starts at a heliocentric distance of about 5 astronomical units (AU). The absolute magnitude of these comets varies from 1.5 to about 9. A small telescope of about 14 inches (e.g. Celestron-14) is sufficient to study some of the events for a long time in a detailed fashion. However, for such studies to have scientific importance, it is necessary that the telescope be backed by a sophisticated back-end instrument. For example, the present author studied Halley's Comet with a Celestron 14-inch telescope equipped with a small imaging Fabry-Perot Spectrometer (Debi Prasad, 1989). The typical problems in cometary science are plasma motion in tails, outbursts and determination of nuclear rotation period. Figure 2 gives a typical example of a cometary tail event. Time-varying processes involving the solar wind's interaction with a cometary plasma tail vary from about 2 to 10 hours. The rotation rates of a cometary nucleus are from a few hours to a few days. Outburst and jet events last for about one to eight hours. Any observing programs should plan the use of small telescopes taking into account these timescales.

SOLAR PHYSICS

One of the main aims of research in solar physics is to understand the properties of solar flares, understanding of which is useful in predicting space weather. An example of a solar flare is given in Figure 3. Solar flares occur because of instabilities in the Corona causing magnetic reconnection at these heights. Consequently, the released energy at these heights propagates downwards and hits the photosphere. This heats the lower atmospheric layers of the Sun and causes spectacular coronal features to be displayed. It is still not known what causes coronal disturbances. Obviously, since the magnetic fields are connected to the solar photosphere, it is likely that the signature of the cause can be traced to this level. Our study shows that pre-flare signatures in terms of flux emergence can be detected at the photospheric and chromospheric layers. An example of such an event is given in Figure 4. Such studies require telescopes of 5-7 inches in size.

SOCIOLOGICAL ASPECTS OF SMALL TELESCOPES

The long-term sustained growth of any nation requires the practice of fundamental science. Astronomy is one of the finest of fundamental sciences, hence it is important to nurture and educate the general public in this subject. From this point of view, it is necessary to build cost-effective research-class telescopes for university grade education. Assuming that about five students opt for a post-graduate course in astronomy, let us analyze the telescope and other peripheral requirements for training them through a degree. The cost of a 12-inch-class telescope with mounts and computer control is about $5,000.00. A Charge Coupled Device (CCD) and a computer would cost about $7,000.00. Other costs such as filters and logistics may run to $3,000.00. Therefore, with a modest budget of about $15,000.00, a good astronomical observatory for university research can be established, which will be suitable for research-class observation. This observatory can be improved gradually by the addition of more advanced back-end instrumentation.

ACKNOWLEDGEMENTS

The author acknowledges financial assistance from the United Nations Development Programme to attend the conference at Toulouse, France.

REFERENCES

Debi Prasad Choudhary, 1989, Ph. D. Thesis, Gujarat University (Physical Research Laboratory)
Diseny, M. J., 1972, Mon. Not. R. Astron. Soc., **160**, 213
Pandey, A. K., Mahra, H. S., Mohan, V., 1984, Bull. Astron. Soc. India, 12, 258
Sagar R., Pandey, A. K., Mohan, V., Yadav, R. K. S., Nilakshi Bhattacharya, D., and Castro-Tirado, 1999, A. J., Bull. Astron. Soc. India, 27, 3
Sagar, R., 2000, Current Science, **78**, 1076
Venkatakrishnan, P., Debi Prasad Choudhary, Ashok Ambastha and Sushant Tripathy, 2000, Multi Aperture Solar Telescope (a proposal for modern ground based facility for solar observations).
Warner, B., 1986, IAU Symp., 1972, 3

FIGURES

Figure: 1. The six GONG stations.

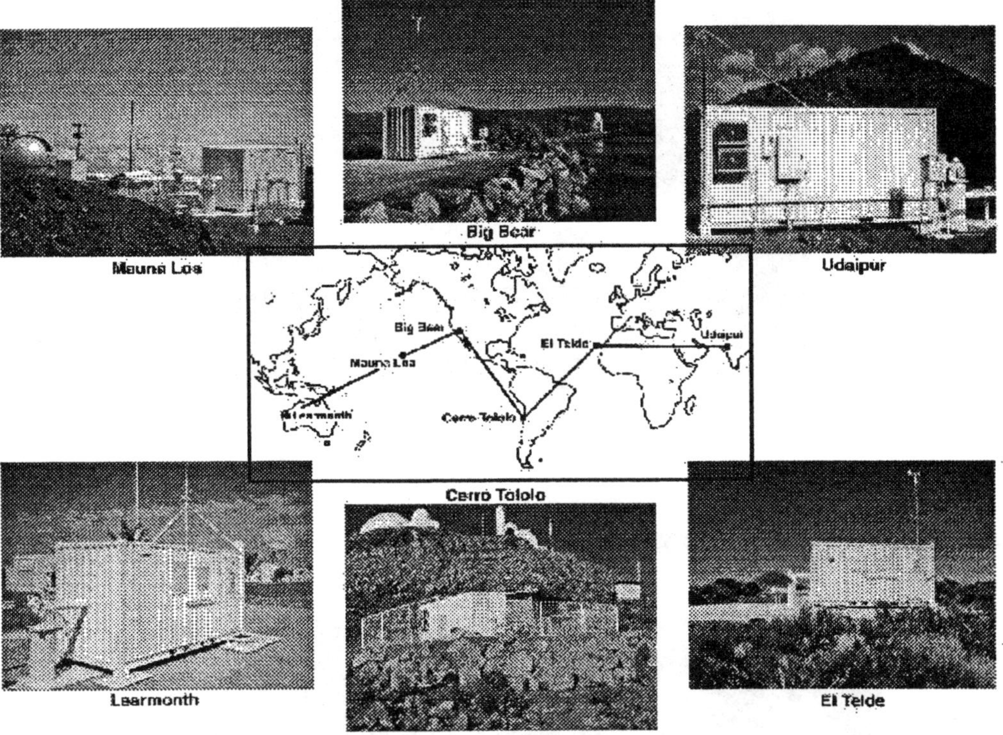

Figure: 2. Typical example of a cometary tail event.

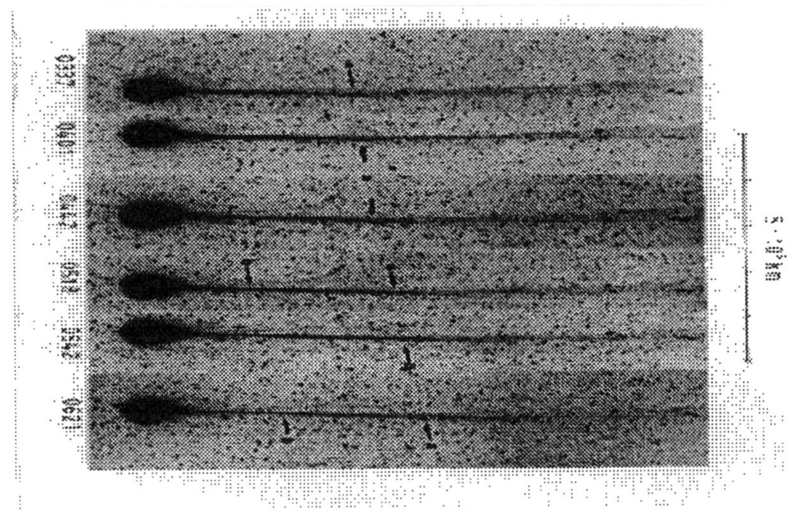

Figure: 3. Typical example of a solar flare:

Figure: 4. Example of a flux emergence prior to the flare.

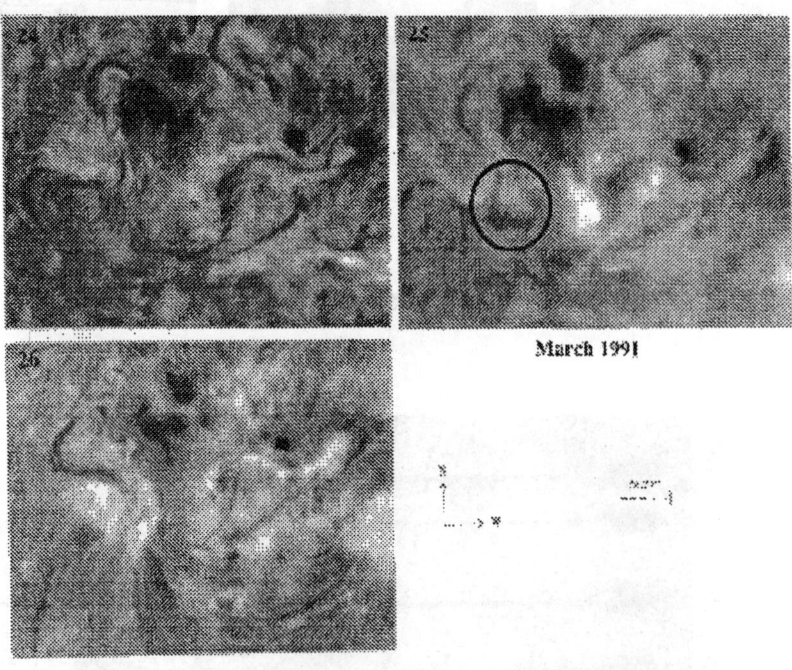

BASIC SPACE SCIENCE IN ETHIOPIA: A STATUS REPORT[*]

Solomon Zewde
Director, National Scientific Equipment Center
Ethiopian Science and Technology Commission
Addis Ababa, Ethiopia

ABSTRACT

The working definition of basic space science and its relevance in answering mankind's millennia-old quests on its origins, evolution and future and on the cosmos in general is linked in this paper to a status report on basic space science in Ethiopia. Accordingly, historical perspectives of ancient Ethiopia on the subject, the modern educational system, the place accorded to basic space science in the modern educational system and related facts are highlighted. Pooling and sharing of knowledge and resources at the national, sub-regional, regional and international levels are also considered and a strategic solution for the creation of an enabling environment for basic space science development in Ethiopia is indicated. Finally, a concise set of recommendations is presented to indicate possible future directions. The goal is therefore the initiation of a detailed basic space science planning schedule at various levels in the country's overall list of activities.

1. INTRODUCTION

According to the Big Bang Theory, matter itself came into being some 15 billion years ago in the aftermath of the Big Bang, the event when space and time began. Then, we are told, followed a formless sea of particles that led to the structure of the Universe and gave birth to the galaxies and stars. Our home planet is believed to have condensed from a colossal supernova explosion of a certain massive star that lived a short life of violent intensity. A chain of events succeeded this. Gaia's Odyssey has it that the Supercontinent "Gondwanaland" separated into South America and Africa about 100 million years ago while about 30 million years ago, Arabia broke away from Africa eventually creating the Red Sea. Jumping forward in time, it is recorded that, 3.5 million years ago, small-brained, bipedal, hominids know as australopithocines, or "Southern Apes" appeared in eastern Africa. More precisely, paleoanthropological research conducted by an American-French Team confirmed in 1974 that the lady internationally known as "Lucy" or *Australopithecus afarensis* ("Dinkinesh" in Amharic, meaning "you are wonderful") lived 3.2 million years ago at a place called Hadar in the Ethiopian Section of the Great Rift Valley of Africa.

---

[*] *This paper was presented at the "Ninth United Nations/European Space Agency Workshop on Basic Space Science", held from 27 to 30 June 2000 in Toulouse, France, and does not necessarily reflect the views of the United Nations.*

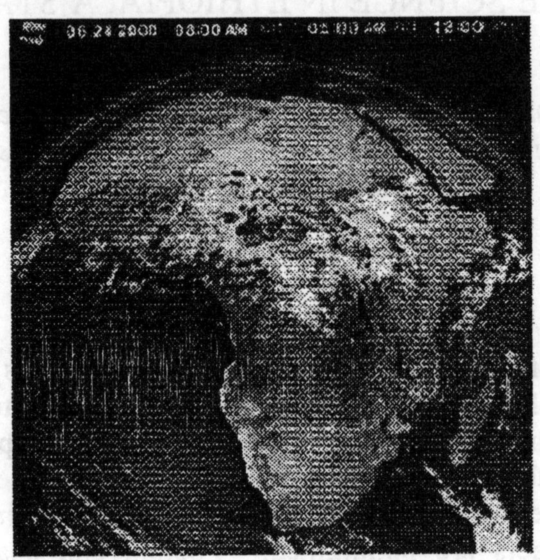

Many people believe the Great Rift Valley might have been the original home of humanity, an idea that accords with the historic speculation of Charles Darwin, in his 1859 Origin of Species, that humanity originated in Africa.

Further archaeological research in recent years reveals that inhabitants of many parts of what is now Ethiopia had embarked, by around 10,000 B.C. (i.e. 12,000 years ago), on artistic activity. Rock paintings from that time have been found in many parts of the region, most notably in Hararghe, Gamo Gofa and Tigray. Some of these pictures in caves depict the milking of cattle, while others feature the use of bows and arrows, spears and shields. There are also representations of humped and humpless cattle, goats, lions and elephants.

Among the foodstuffs at least three are known to be native to the country: the widely diffused and highly prized grain teff (*Eragrostis teff*); the oil seed nug (*Guizotia abyssinica*); and the "false ensete" (*Eduulis edule*). All three probably began to be cultivated at about the same time, perhaps by or before the 5th millennium B.C. (i.e. 7,000 years ago).

Ethiopia is also the source of coffee and there are anthropologists who believe the Ethiopian alphabet is the source of the many of the world's alphabets, including Latin, Greek and Arabic.

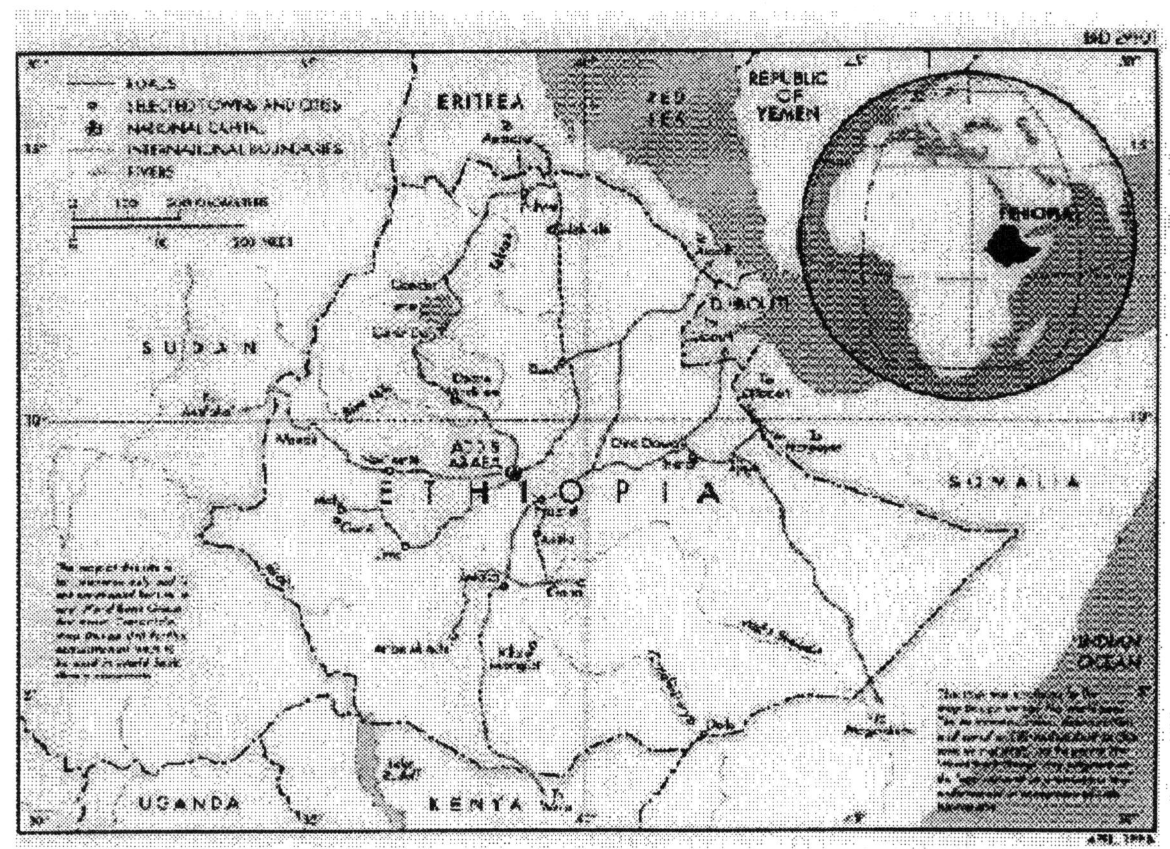

A cradle of mankind and located at 3-15°N in an extent as large as France and Spain combined, does Ethiopia therefore have any space science and astronomical heritage? The paper will return to this topic shortly. For the moment, however, attention will be given to the definition and scope of basic space science itself, to tie things together.

Outer space, or simply "space", is everything from the middle levels of the Earth's atmosphere (about 60 km above sea level) to the edge of the Universe several billion light years away. Basic space science, as a consequence, is the fundamental domain of knowledge about this almost endless expanse: the Universe, and its origins, evolution and destiny. For working purposes, basic space science may be defined to encompass the four broad areas of (credit: Dr. Peter Martin, Republic of South Africa):

- Astronomy and astrophysics;
- The solar-terrestrial interaction and its influence on terrestrial climate;
- Planetary and atmospheric studies; and
- The origin of life and exobiology.

Apart from astronomy and astrophysics, basic space science, as a fully fledged science, makes use of many other sciences that include:

- The science of astronautics;
- Geophysics;
- Meteorology;
- Space physiology (space medicine);
- Space chemistry;
- Space biology; and
- Other related fields.

It is simply amazing in all senses that since the birth of the space age on 4 October 1957, with the launch of the first man-made satellite, Sputnik I, land and space based tools have become widely available for the study of space science. The whole effort in this respect may be characterised as being part of a grand quest to understand our cosmic origins and destiny, and how these are linked by the cycles of evolution. Is the Universe expanding, contracting or remaining the same size? What about its shape and density? What are the effects of these parameters on the overall destiny of the Cosmos and life? Do water and life as we know them on Planet Earth exist on Mars or in any other part of the Universe? And so on and so forth – there is no end to the questions. In any case, these are some of the fundamental questions a dedicated study of space science intends to cover.

And, what is more, the above questions invite one to consider still more questions. Has Ethiopia contributed in any way towards answering these questions? What are the future prospects in this regard, if any? In the following sections, direct information on these and related issues shall be presented.

2.   ANCIENT ETHIOPIA: HISTORICAL PERSPECTIVES

In order to understand his relationship to his gods, ancient man turned to a fixed system of observations, many of which were extraordinarily ingenious and accurate. This was the situation in the ancient world and it was the same under the Ethiopian skies. In time, control of such activity became concentrated in the hands of the priesthood as can be verified through various records. The result was that, instead of developing as a scientific practice, astronomy remained a superstitious cult.

As for the signs of the Zodiac, twelve stars could be identified in Ethiopian astrology. In accordance with the established belief in ancient Ethiopia that all matter was composed of the four building blocks fire, water, earth and air, the 12 stars were divided among these four groups, with each group consisting of three stars. As the cult was developed further, the fate of a person was related to these stars through a process of ascribing numbers 0-12 to the letters appearing in the person's and his mother's names. The value 12 was subtracted each time the sum of the numbers obtained exceeded 12. The resulting single number was considered the star of the person in question. Various interpretations were given to the various stars. Relationships between the stars also existed; for instance, marriage between two fires was not recommended; in this case the wife was advised to change her name so that she would obtain a better star.

Astrology in ancient Ethiopia also had other aspects including the claim of investigating events.

In local terminology, the cult of astrology is known as "Mesthaf Geletsa", which means "Investigating the Books" while the occult practitioner who specialises in the trade is named "Metshaf Gelach" meaning "Expert in Investigating the Books". There is practically nothing that the *Metshaf Gelach*s does not claim to know. The fact remains, however, that astronomy could not evolve out of the fetters of such a situation.

Apart from the practice of astrology, oral traditions testify to the fact that star gazing ("Kokeb Kotera" = star counting) was practised in ancient Ethiopia; follow the next popular lines (presented in the Latin alphabet):

<u>Song 1</u>:

Abeba ayeh woi,	Lemlem
Abeba ayeh woi,	Lemlem
Abeba sisha,	Lemlem
Tisha le Tisha,	Lemlem
Libelagn neber,	Lemlem
Ya tikur wusha,	Lemlem

Abeba yebir muday Kolel bey, Abeba yebir muday Kolel bey

Balinjeroche / Kumu betera
Enchet sebre / bet eskisera
Inquan betina / yelegnim atir
Edej adralehu / Kokeb sikotir
Kokeb Kotre / sigeba ebete
Tikotagnalech / yenjera enate

<u>Song 2</u>

Aleh hoi, Aleh hoi
Gereden alayeh woi
Gerede Mariam sima
Birkuma teshekima
Birkuma yedagete

Chereka Dimbuloka/   Atse bet Gebach awka
Atse bet yalut lijoch/   Inkirdad fetagiwoch
Fetegu fetategu/ Bekulbit askemetu
Kulbitua bitiseber/ Bewancha gelebetu
Aja kolo Sinde kolo/   Yichin tilesh yichin tolo

Focussing on the central and relevant ideas of the above songs, one sees that, in the first song, the singer passes the night outside her home observing the stars, at the end of which, when back at home, she is not on good terms with her stepmother.

In the second song, the singer, a female this time also, sings about the lovely Moon entering the King's home. Naturally, associating the Moon with the King, the most respected person in ancient Ethiopia, is indicative of the fact that Ethiopians really loved astronomy.

On the other hand, the first song is sung once every year, on the occasion of the Ethiopian New Year in September, while the second song is sung quite commonly by young children. These songs have been transmitted by word of mouth from generation to generation.

To add some information to the above, we next consider the story of Andromeda. Ethiopia is quoted in Greek mythology through the chained lady, Andromeda, daughter of King Cephus of Ethiopia and his wife Queen Cassiopeia. Cassiopeia, we are told, boasted that her beauty, and that of her daughter, exceeded the beauty of the sea nymphs. This angered Poseidon, god of the sea, who sent the mighty whale Cetus to destroy the Ethiopian coast. The only way to calm the monster was to sacrifice Cepheus' lovely daughter to it.

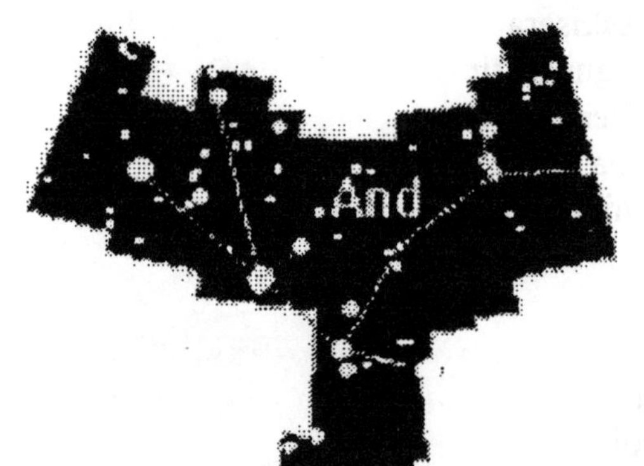

Neighboring Constellations:
[Cas | Lac | Peg | Per | Psc | Tri]

After a long fight, the parents finally decided to chain Andromeda to a rock in the ocean. Luckily, just at this moment, the hero Perseus came to Andromeda's rescue. It is said that Perseus changed the whale into a rock in the sea. Cephus was happy and gave Perseus Andromeda's hand and a kingdom of his own. The gods were so pleased that they granted a place among the stars to the actors of this drama ...

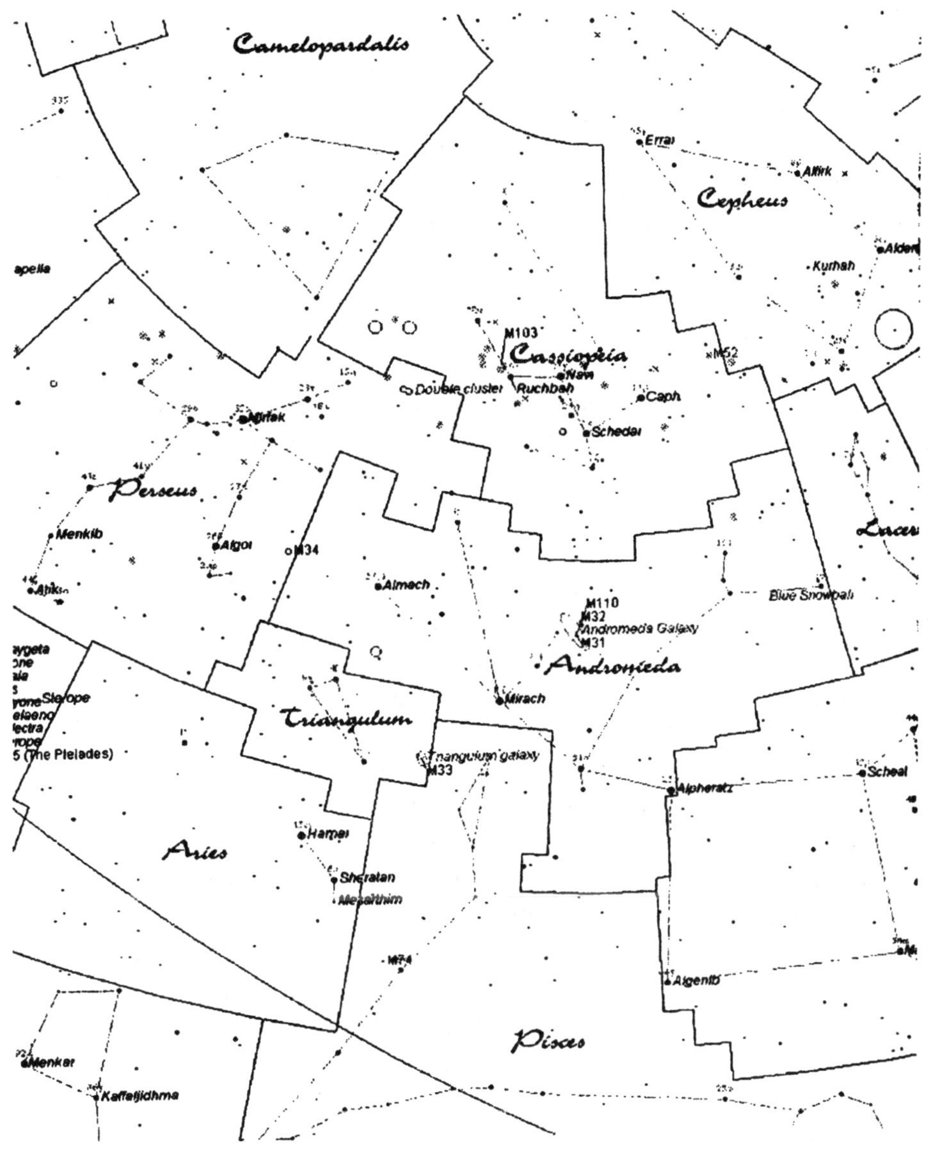

Andromeda's declination is 22°N-53°N. The constellation is visible in the northern hemisphere from June to April and the center culminates at midnight around the 30th of September. Constellations neighbouring Andromeda include Cassiopeia, Lactera, Pegasus, Pisces, Triangulum and Perseus. From Cephus, Andromeda is distant through Lactera and Cassiopeia. Andromeda is abbreviated as And.

Therefore, in summary to this section, one sees some form of astronomical experience prevalent in ancient Ethiopia, full details of which are fairly left for study and research by anthropologists and historians.

3. MODERN EDUCATION IN ETHIOPIA AND THE PLACE OF BASIC SPACE SCIENCE

According to a recent report by Noel Okoth of the All African News Agency, the literacy level in Ethiopia has increased from less than 35 per cent in 1990 to about 60 percent. This is no mean task for a country frequently hit by several evils of progress. Characteristically in Ethiopia, and indeed in many other countries in sub-Saharan Africa, the crucial challenge facing education in general and education in basic space science in particular is linked to limitations in resources. The choice has usually been whether teaching the alphabet or conducting scientific research should take precedence over building roads and infrastructure or providing basic health services to children and expectant mothers threatened by high morbidity and mortality. Even taking into account these harsh choices that must be made, Ethiopia, again according to Noel Okoth, is spending 4% of its Gross National Product (GNP) on education. The share of basic space science in this allocation of resources, however, is something that needs to be revisited and reconsidered.

The new educational system for Grades 1-12 is sub-divided into four levels: Elementary (grades 1-4); Junior High School (grades 6-8); High School (grades 9-10); and Pre-college (grades 11-12). Enrolment ratios vary from 42% for Grades 1-8 to 8.9% for Grades 9-12. A close look at courses taught in these grades shows that a "space" topic is introduced to the Ethiopian student for the first and last time in Grade 6, and as a part of two subjects at the same time: the Science and Social Studies classes. One can therefore sense the need for more continuity, avoidance of repetition and introduction of space science topics at earlier levels. Topics related to history and the structure of the Universe, identification and location of constellations, the history of flight and the Solar System should be considered for inclusion in the curriculum for grades 1-12, along with discussion of the cycles and effects of the Moon on the Earth, terrestrial life and human activities. While some of these topics might have to be lumped into comprehension classes in language courses because of time constraints, establishing astronomical clubs in schools is another possibility.

Tertiary level education also needs attention. The various universities and colleges in Ethiopia have no full-fledged space science programs. There are some related courses distributed among a number of fields. The Addis Ababa University, the largest University in Ethiopia, for instance, teaches various physics, meteorology, geology, mathematics, biology and chemistry courses as part of its overall curriculum. However, specific courses like celestial mechanics, astronomy, astrophysics and astrophysiology are lacking. The trend is the same at both the undergraduate and graduate levels. The department of physics may be considered more closely. In this department, the BSc program in physics involves a minimum of 60 credits for students in the education team. Physics courses at the undergraduate level fall into one of the following eight areas:

- General physics and lass;
- Mechanics;
- Thermodynamics and statistical physics;
- Quantum physics;
- Electrodynamics and electronics;
- Nuclear physics; and
- Topics in physics (e.g., senior project).

In course coding schemes, a three level coding system, in which the middle number refers to one of the eight areas mentioned above, is utilised. The first digit of the scheme indicates the year a course is offered while the last digit, which may be either an odd or even number, refers to the semester. All in all, there are 23 physics and two electronics courses in the undergraduate physics program. In addition to compulsory sets of laboratory exercises, the student also takes courses in Mathematics, Biology, Chemistry, English, Geography, Economics, Geology, Physical Education and Pedagogical Science. The central theme of the program is to develop a well-rounded personality and education.

The description of courses shows that a wide range of topics is covered. In particular, the following topics have much in common with a basic space science education program:

- Gravitation and Planetary Motion;
- Relativistic Momentum and Energy;
- Thermal Radiation;
- Classical Mechanics;
- Quantum Statistics;
- Optics; and
- Nuclear Physics.

The above wealth of topics maybe coupled with Celestial Mechanics, Cosmic Magnetic Fields, High Energy Astrophysics and related topics to launch a successful and strong basic space science education program at the university level. The foundations already exist; one needs only to build upon them.

The MSc Program in Physics at the Addis Ababa University involves a minimum of 24 credits of coursework and a research project leading to a Thesis, which must be defended successfully. Courses offered are in the following areas:

- Advanced Physics Laboratory;
- Classical Mechanics;
- Mathematical Methods of Physics;
- Electromagnetic Theory;
- Statistical Mechanics;

- Quantum Mechanics;
- Solid State Physics;
- Semiconductor Physics;
- Atomic and Molecular Physics;
- Nuclear Physics; and
- Laser physics.

MSc projects in Nuclear Physics, Solid State Physics, Quantum Optics, Laser Physics and Polymer Physics are offered as choices for research work, as there exists various scientific equipment such as a spectrophotometer, detectors, analysers, neutron sources, multimeters and related hardware to support the research work. It is also good news that, not unlike the undergraduate courses, there are also space science topics contained in the above areas of study.

The existing strong programs in Chemistry, Biology, Geology and Mathematics at the Addis Ababa University may also be linked to a Basic Space Science Education Program. It would be a good practice to make use of existing resources.

The Addis Ababa University also owns Geophysical and Astronomical Observatories at its Science Faculty premises. The Geophysical Observatory is involved in the early detection and monitoring of seismic and other geophysical phenomena. The Astronomical Observatory at the Science Faculty of the Addis Ababa University is an old and small observatory used for teaching purposes. There is another Astronomical Observatory near the city of Debre Zeit but this is no longer functional. In light of current developments, it is very important and still not too late to build bigger and modern astronomical observatories for Ethiopia in various localities. On a much smaller scale, space science and astronomical tools like planetaria, charge coupled device (CCD) cameras, binoculars, optical and infrared equipment and telescopes are necessary in the immediate short-term at various levels from lower grades to the post-graduate levels, for educating students and the public and to enable further scientific studies in space science and astronomy.

4.   THE NEED FOR COLLABORATIVE VENTURES

As shown so far, there is the cultural background and practical need to develop space sciences in Ethiopia. This requires collaborative efforts of many parties at national, sub-regional and international levels. The Ethiopian Science and Technology Commission, the Ministry of Education, the Addis Ababa University and other organisations may be identified for this purpose. Departments such as those of Electrical Engineering, Physics, Mechanical Engineering, and the Geophysical Observatory, all of which are part of the Addis Ababa University, should take the initiative to co-operate in several areas, including development and implementation of a project for an indigenously constructed satellite that would be built by the staff and students in these departments in co-operation with external assistance. The University of Stellenbosch in the Republic of South Africa may serve as a model and an example to follow. This project, apart from being useful for resource imaging and data relay, would also be instrumental in the introduction of satellite engineering and operation skills into Ethiopian Academia along much the same route the University of Stellenbosch chose to follow.

At the sub-regional level, there is an already functional Deputy Co-ordination for East Africa of the Working Group for Space Science in Africa (WGSSA). The Co-ordination should be strengthened to realise scientific integration among the concerned countries.

Co-operation in Space Science and Astronomy at the African level among the various African professionals has been encouraged, particularly since establishment of the WGSSA in September 1996 in Bonn. The WGSSA has so far published four issues of the journal "African Skies/Cieux Africains" (May 1997, April 1998, January 1999 and December 1999, numbers 1 to 4, respectively). Ethiopia's continuous participation in WGSSA is secured.

Co-operation at the international level to promote development of basic space science and astronomy in Ethiopia is particularly important. The international scientific community has made use of various occasions to perform intensive studies of the Earth and its environment. Ethiopia's inclusion in a space project, either on a short- or long-term basis will create further opportunities to develop basic space science worldwide. In doing so, Ethiopia will not only obtain the impetus to develop the field in its region, but will also actively contribute towards the global heritage of scientific knowledge, practice and discoveries.

5.   THE WAY FORWARD – SUMMARY AND RECOMMENDATIONS

Realising the importance of the cultural heritage, transferred to the current Ethiopian generation through various means that include oral traditions, in the creation of a conducive atmosphere for the development of Basic Space Science and Astronomy in Ethiopia;

Understanding the low level of development of basic space science and astronomy in that country;

Appreciating the need to use existing resources at the Addis Ababa University and elsewhere locally;

Cognisant of the role played by international co-operative ventures; and,

Recognising the need to establish the Ethiopian society of space science and astronomy;

It is recommended that:

5.1   The Ministry of Culture and Information, astronomers, anthropologists and other scholars and parties should collect, compile and do research work on the astronomical heritage of Ethiopia to arrive at a final reference literature on the topic.

5.2   The Ministry of Education should consider opportunities of curriculum review to introduce the student to a strong basic space science and astronomy background in grades 1-12.

5.3   Ethiopia should take the initiative to participate in sub-regional, regional and international co-operation in the fields of basic space science and astronomy on a wider scale.

5.4   The Addis Ababa University should likewise revise its curriculum and introduce basic space science and astronomy, both as courses and fully-fledged programs of study at the undergraduate and postgraduate levels.

5.5   The Government of Ethiopia, through the Ethiopian Science and Technology Commission and/or any other office competent for this purpose, should take the initiative to develop space science and astronomy by allocation to this purpose some of the already meagre resources.

5.6   Professionals and interested parties should work towards establishment of an Ethiopian Society of Space Science and Astronomy (ESSSA).

5.7   Schools should encourage young pupils to join astronomy clubs first in their respective compounds and ultimately to form a joint national council.

5.8   The International Space Science and Astronomy Community should assist Ethiopia in various respects towards promotion and sustainability of space science and astronomy in Ethiopia.

ACKNOWLEDGEMENTS

The author has used several publications in the preparation of this document and would like to express his gratitude to the following persons and organisations for the use of the literature mentioned alongside their names:

1.   Dr. Mulugeta Bekelle, Department Head, Physics Department, Faculty of Science, Addis Ababa University (for making available course descriptions for the under graduate and graduate program).

2.   Professor Ermias Dagne, Professor of Chemistry, Faculty of Science, Addis Ababa University (for making available the courses of study of the Department).

3.   Dr. Peter Martinez, South African Astronomical Observatory, PO Box 9, Observatory 7935, South Africa (for the use of his paper "Basic Space Science in Africa")

4.   Michel Blanc and Monique Querci, Observatoire Midi-Pyrénées, 14, av. E. Belin, Toulouse, France (for the use of their paper "Observatoire Midi-Pyrénées")

5.   "Gaia's Odyssey", Internet material

6.   Abebe Kebede, EDLA Ethiopian Distance Learning Association (for the use of his material entitled " Ethiopian Astronomical Society")

7.   Donat G. Wentzel, Professor Emeritus, University of Maryland written at the Internet site http://www.seas.columbia.edu/rah297/un-esa/astrophysics/contents.htm (for the use of "Astrophysics of University Courses")

8.   Advanced Learning Systems for the material in the Web site http://www.tarmaced.com/cg-earth-space-science-htm (title of material: "ATLS Curriculum Vitae")

9.   David N. Spergel, Microwave Anisotropy probe (for the use of "The Big Bang Theory" – Internet material)

10.   The NASA Web site, for various material on Space Science

11.   WGSSA, Working Group for Space Science for Africa, "Newsflash".

12.   Prof. Dr. Hamid M. Al-Naimy, Instrument of astronomy and Space Sciences, Al al-bayt University, P.O.Box 130040, Magrag, Jordan, "Jordan develops Astronomy at the Research and Popular level"

13.   Dr Richard Pankhurst in Selamta volume 17, number 2, "Lucy and Ethiopia in the ancient world"

14.   Demissu Gemeda, Associate professor at the Department of mathematics, Faculty of Science Addis Ababa University, "Higher Education and Research in the Basic Sciences in Ethiopia", Addis Ababa, January 1999.

15.   Demissu Gemeda, Associate professor at the Department of mathematics, Faculty of Science Addis Ababa University, "Status of Basic Science Education and Research in Ethiopia", Addis Ababa, (1999?)

16.   Demissu Gemeda (PhD) and Yismaw Alemu (PhD), "Study Report on Tertiary Mathematics Education in Ethiopia", Addis Ababa, August 1998.

17. Bernard Nies, http://www.astroinfo.ch/dsc/And-Myth.e.html, "Astro Info Deep-sky corner/Andromeda", last modified 22 January 1998.

18. Johannes Andersen, International Astronomical Union, 98bis,blvd Arago, F-75014 Paris France, "The IAV and its role in the world wide development of astronomy", Working Group, Sun Jan 17, 1999.

19. A. Hababyanna and G. Muhyeme, Physics Department, University of Zambia, "Basic Space Science Education and Research in Zambia", WGSSA, 24 February 2000.

20. M.G. Iziomon, Meteorological Institute, University of Freiburg, Werderring 10, D-79085 Freiburg, Germany and Department of physics, University of Illorin, P.M.B. 1515, Illorin, Nigeria, "Promoting space science education in Africa: practical initiatives", WGSSA, February 24 2000.

21. WGSSA, "News/Nouvelles", 25 February 2000.

22. United Nations Office for Outer Space Affairs (OOSA) (for using their web site and printed matter thereof).

23. Mr Noel Okoth, All Africa News Agency, "Poor resources hinder academic growth in Ethiopia", date not mentioned.

THE ASTRONOMICAL OBSERVATORY AT UNIVERSIDAD NACIONAL DE ASUNCION, PARAGUAY, A RESULT OF COOPERATIVE EFFORTS OF ASTRONOMERS FROM THE NATIONAL OBSERVATORY AT MITAKA, JAPAN, INTERNATIONAL ASTRONOMICAL UNION COMMITTEE #46, THE UNITED NATIONS OFFICE FOR OUTER SPACE AFFAIRS AND THE GOVERNMENT OF PARAGUAY[*]

A. E. Troche-Boggino
Observatorio Astronomico, Facultad Politécnica
Universidad Nacional de Asunción

I   BACKGROUND

Paraguay has one of the earliest astronomical observatories in South America, built by Buenaventura Suárez. Suárez is an example of a self-made astronomer in colonial Paraguay. He built his own appropriate instruments and applied his own rules for computation. He obtained data from his own observations. He published his data in a book and wrote reports to other scientists around the world.

In the town of San Cosme y Damian, one of the thirty Jesuit communities for Guarani Indians in the Great Province of Paraguay, lived an excellent 18th century astronomer, Father Buenaventura Suárez. Father Suárez was the first native astronomer from the southern regions of South America. He was born in Santa Fe on September 3rd, 1678, and studied at Cordoba. He did most of his work in San Cosme y Damian, from the early 18th century until his death in Santa Maria, on August 24th, 1750. San Cosme y Damian is in present-day Paraguay (the other three cities are in Argentina).

With the help of local native artisans, Suárez built various astronomical instruments: Kepler-type refractors with lenses polished from local crystalline rocks; sundials; a quadrant with degrees divided into minutes; and a pendulum clock with minutes and seconds marked. In 1743, the Jesuit Order provided him with telescopes (with focal lengths ranging from 2.2 to 6.5 m) and two Martirion clocks, imported from England. For 13 years, Suárez accurately observed eclipses of Jupiter's satellites. He also observed eclipses of the Sun and the Moon. He made determinations of the longitudes (as meridian differences) and latitudes of San Cosme y Damian and all Jesuit mission towns. He corresponded with various famous astronomers of his time.

---

[*] *This paper was presented at the "Ninth United Nations/European Space Agency Workshop on Basic Space Science", held from 27 to 30 June 2000 in Toulouse, France, and does not necessarily reflect the views of the United Nations.*

In 1743, he printed the first edition of "Lunario de un Siglo", an astronomical calendar containing the phases of the Moon, solar and lunar eclipses, church festivities and geographic coordinates of 70 cities. This book was re-issued four times until 1858.

The Jesuit priests and brothers were expelled from Spain and all of its colonies around 1767. Time and military expeditions took their toll. All of Father Suárez's instruments have been lost with the exception of a sundial at San Cosme and Damian, which now serves as a lonely testament to this exceptional man. We would like to find an original copy of Suarez's "Lunario de un Siglo", to re-edit it as a valuable part of our historical heritage. We seek your help looking in libraries for this elusive book.

II   STARTING AN ASTRONOMY PROGRAM

I have been active in astronomy education since 1974. Astronomy was then taught only sparsely in Paraguay. I found opportunities to train high school teachers; a new program for secondary schools was in a trial stage. Basic Astronomy then became part of high school Natural Science and Mathematics courses. An elective astronomy course for junior university physicists was also placed on the curriculum.

Encouraging International Astronomical Union (IAU) International Schools for Young Astronomers (ISYA's) were held in Argentina (1974) and Brazil (1977 and 1995). Contact with IAU Commission #46 was soon established. I was advised on how to start a supporting framework for my country. Some IAU astronomers visited us. A proposal for an IAU Visiting Lecturers Project (VLP) in Paraguay appeared on the horizon, became real in 1988 and ran until 1994.

Equipment from IAU-TT and a telescope were borrowed for hands-on astronomy. Some results: three students went to Brazil-ISYA (1995), two others started working as astronomers abroad and others became university and high school instructors. Experience demonstrated the need to find jobs for these young astronomers. This led to the additional project to develop an Astronomy Center around an Astronomical Observatory.

III   NATURE HELPS: THE INTERNATIONAL SOLAR ECLIPSE FORUM 1994

A group of professors, astronomers and students from Meisei University of Tokyo visited Paraguay to observe a total solar eclipse. They came under the leadership of Dr. Eijiro Hiei and Mr. N. Takahashi. Two days before the solar eclipse, Meisei University and the National University of Asuncion held an historical international forum in the National University at San Lorenzo Campus. Almost 300 people attended. As far as I know, this was the first international meeting on astronomy ever held in Paraguay.

Dr. Eijiro Hiei, a National Astronomical Observatory at Mitaka emeritus professor and solar astronomer, suggested that we apply to the Government of Japan for Overseas Development Assistance (ODA) to obtain a telescope and its accessories.

So we applied twice, first in 1996 and again in 1997, to the Cultural Project Cooperation of the Japanese Government to Paraguay, in the areas of education and culture, through the Embassy of Japan in Asuncion.

Visits of Dr. Percy and Dr. Kitamura in 1997 and 1998, respectively, were particularly useful to support the project for the astronomical observatory and in particular to extend the project to a "Center for Astronomy in Paraguay".

Early in 1998 we learned that our project was selected for ODA, a record time compared to other countries who made similar applications. Dr. Hans Haubold at the United Nations Office for Outer Space Affairs contacted the Paraguayan Mission to the United Nations in Vienna. It was very opportune. Right on the final day for presenting the project to the Japanese Embassy in Asuncion, our Ministry of External Relations decided to support the telescope project.

By August of 1999, the astronomical instruments were at the customs office at the port of Asuncion.

IV   THE EQUIPMENT DONATED

The 45-cm Cassegrain-type telescope is furnished with two high-technology pieces of equipment: 1) A digital photoelectric photometer with UBV filters – the signal is amplified to a million-times' strength and lead to the recorder and digital printer controlled by a computer; and 2) a Charge Coupled Device (CCD) and image processing equipment.

V   THE OBSERVATORY BUILDING

Advice, good draft charts and photographs from the Bosscha Observatory in Indonesia, the Arthur C. Clark Center for Modern Technology in Sri Lanka and observatories in Thailand and Taiwan Province of China (which also received similar telescopes from Japanese ODA), were generously provided by Dr. B. Hidayat, Eng. de Alwys and Dr. M. Kitamura. These were very useful for designing an observatory building with a moving roof. Careful explanation from Dr. Kitamura and permanent contact with Goto Opt. Co was a lot of help in constructing the observatory building. The telescope and its parts were installed and tested for a month until April 17th 2000.

Thanks to support from the Japan International Cooperation Agency (JICA), a young physicist, Mr. Fredy Doncel, was trained for nearly six months at Nishi-Harima and Beisei Observatories in Japan.

VI GOALS

Teaching of astronomy at any level is very necessary, and all young people should learn about space science.

An astronomical observatory is essential to stimulate observations, for data acquisition, publications and to interact with other astronomical centers abroad. The value of a telescope for attracting interest in space science should also be considered. Students should be ready to go further and get involved in astronomical research or become more involved in related scientific activities like remote sensing, meteorology, computer science and communications. Astronomy is a stimulus for training future scientists and high-level technicians.

The following is from Dr. Kitamura's toast during the inauguration ceremony for the Astronomical Observatory at Universidad Nacional de Asuncion:

"We know that the science of Astronomy began with observations. Nowadays, astronomy and astrophysics need more and more observational data. In this regard, I would like to emphasize that the primary importance of astronomical observation is in the geographical situation rather than the telescope size. Any astronomical object cannot be seen by all the observatories in the world simultaneously, and therefore the observational data that will be obtained in Asuncion should be always new. Nowadays, international cooperation is important in any branch of natural sciences. In astronomy, however, this mutual cooperation among different countries is particularly important. Astronomical observations very often concern time-varying phenomena of objects in the Universe. The International Astronomical Union organizes an international campaign of cooperative observations of these particular astronomical objects . . ."

The astronomical observatory was inaugurated on June 6th, 2000.

PROGRESS WITH THE URUGUAYAN AUTOMATED AND ROBOTIC TELESCOPE[*]

Rubens G. Freire Ferrero
Strasbourg Observatory
11, rue de l'Université
67.000-Strasbourg-FRANCE
freire@cdsxb6.u-strasbg.fr

Amarú Alzogaray Hernández
Pro-tempore Secretariat of the Third Space Conference of the Americas
Ministry of Foreign Affairs
Montevideo-URUGUAY
amaru@mrree.gub.uy

ABSTRACT

Project TAURO aims to install a mid-sized automated and robotic telescope in Uruguay, as one step of a more ambitious plan for educational development and technical training in Uruguay. Here, we present the state of advancement of this project.

INTRODUCTION

Astronomy provides a general and captivating approach to Reality, and it can easily open doors to other sciences such as general physics, meteorology and biology and make calculus and mathematics more accessible. But today, astronomy is a Big Science, and to explain astronomical facts it is necessary to have at one's disposal at least a few clever instruments to draw and maintain the attention of students and the public. To pursue these goals, modern information techniques must also be used in parallel to obtain and complete data sets, in particular astronomical images, sky charts and bibliography. With the aim of developing an ambitious plan regarding education at low, intermediate and high levels, we are developing project TAURO, consisting of an automated and robotic telescope to be installed in Uruguay.

The first presentations of this project were made in early 1994 in Montevideo and the project was then submitted to the *Third Space Conference of the Americas* (Punta del Este, Uruguay) in November 1996 and to the *Regional Preparatory Conference for the Third United Nations Conference on the Exploration and Peaceful Uses of Outer Space (UNISPACE III)* (Concepción, Chile) in October 1998.

---

[*] *This paper was presented at the "Ninth United Nations/European Space Agency Workshop on Basic Space Science", held from 27 to 30 June 2000 in Toulouse, France, and does not necessarily reflect the views of the United Nations.*

INTRODUCING URUGUAY: A SUMMARY

The real name of Uruguay is *República Oriental del Uruguay* (the Eastern Republic of Uruguay). Uruguay is a small country in relation to others in Latin America, but it has a "European size": a third that of France or Spain, three-fifths of Italy and double that of Portugal.

- Surface area: 176,215 km^2
- Latitude: between 30° and 35°
- Longitude: between 53° and 58° W
- Population: 3.5 million inhabitants
- Capital city: Montevideo, with 1.8 million inhabitants

The highest peak of the country is the *Cerro Catedral*, 513 m in height, surrounded by other hills of similar height such as *Sierra de los Caracoles* and *Sierra de las Ánimas*.

The mean climate characteristics are:

- Temperate – the mean annual temperature is 17° C with some variability, very similar to that of European Mediterranean countries;
- Rainfall – from 900 mm/year in the south-west to 1300 mm/year in the north-west;
- Mean annual sunshine – from 2300 hours/year in the east to 2550 hours/year in the west;
- Mean cloudiness (in eighths of covered sky) – from 4 in the north to 5.5 in the south;
- Annual mean relative humidity – from 71% in the north to 80% in the south;
- Mean wind speed – from 15 km/hour in the north to 25 km/hour in the south, with a maximum of 200 km/hour.

THE PRESENT STATUS OF ASTRONOMY IN URUGUAY

An astronomy course is still taught at the secondary level but in recent years there have been several attempts to cancel it. At the University, the Department of Astronomy is in charge of the *Licenciatura de Astronomía* (5 years) and also of a *Maestrado de Astronomía* (2 additional years). A group of active amateurs contribute to public activities and popularization.

The present astronomical resources are situated in the Montevideo district at the "Los Molinos" (the wind-mills) Observatory (low-altitude), where there are:
- A 35-cm telescope (Broadhurst Clarkson & Fuller) donated by the British Council;
- Two 25-cm amateur telescopes;
- Charge Coupled Device (CCD) cameras.

THE PRINCIPAL GOALS OF PROJECT TAURO

To establish the basis for an Astronomical Observatory in Uruguay:
- Far away from population centres;
- Using an automated and robotic telescope to allow remote control;
- With some independence from governmental and private organizations.

To develop complementary action for education and research, using a non-traditional approach:
- Distant access to observe, to retrieve and to analyze astronomical images or spectra for educational and research purposes;
- The use of an automated and computerized system to contribute to technical education;
- Serving to promote basic space science;
- Acting to popularize the scientific approach to reality.

To take advantage of a profitable climate in other regions of Uruguay in relation to:
- Economy (local capabilities);
- Tourism (foreign contributions);
- Education (regional).

A minimal instrumental configuration could be:
- A Cassegrain telescope, minimum 60 cm in diameter (maximum 1 m), approx. F/5;
- A CCD camera for imaging and to record spectra;
- A CCD camera for the guidance telescope;
- Electronic devices for automated control;
- A computer system allowing for local and remote control, to record, to stock, to implement some data processing and to distribute digital images and data.

Complementary software and hardware would be needed to take advantage of available computer resources for obtaining and to reduce astronomical images, carry out some data processing, establish rapid links with other net sites and to enable preparation of didactic educational material.

For research purposes, the instruments could be used:
- To study active and variable stars;
- To search for novas or supernovas;
- To study asteroids and comets; and
- To search for Near Earth Objects;

using photometry and spectroscopy techniques, in coordinated international campaigns.

FIRST STEPS ACCOMPLISHED

- Choice of Maldonado province (150 km east of Montevideo);
- Contacts with local authorities in Maldonado province;

- Preliminary choice of the site;
- Contacts with the National Meteorological Service;
- Contacts with ANTEL (National Telephonic Administration) to provide support to install telephonic lines (or transmitter and receiver).

RESULTS OBTAINED

<u>From the Maldonado Mayor's Administration:</u>

Oral agreement about the importance of the project for local economy and tourism and for education in general at the regional level.

Oral agreement on local financial support for infrastructure, including:
- Transfer of a common piece of land;
- Extension, construction or improvement of roads;
- Help to install electricity, water and sanitation;
- Construction of the roll-off building to shelter the telescope;
- Construction of the Observatory building;
- Financial support to some administrative and technical staff.

<u>From the National Meteorological Service:</u>

Oral agreement to install an automated weather station (AWS), which is now out of normal service.

NEXT STEPS

On the chosen site:
- Installation of an AWS;
- Planning several nocturnal campaigns at different seasons to analyze the astronomical quality (seeing) of the sky;
- To carry out a preliminary topographic study to draw up plans for construction of the building.

Organization:
- Creation of a Foundation to support and supervise the installation of the telescope and the Observatory organization;
- An Official Local Committee, delegated by the Foundation, will organize and develop contacts to accomplish the principal goal;
- External members of the Foundation will help to obtain international support through personal and official contacts.

The advantages of a foundation are:
- Versatility of administration and management;

- Quick decisions;
- Apolitical organization;
- Direct collaboration and contacts;
- Search for funds and financial support from multiple sources.

We will take into account the important collaboration with Local and National Partners:
- Province of Maldonado;
- Ministry of Foreign Affairs – Pro-tempore Secretary of the Third Space Conference of the Americas;
- Ministry of Education and Culture – University of Montevideo;
  - Faculty of Sciences
  - Institute of Physics
  - Department of Astronomy
  - Department of Meteorology
- Ministry of National Defense;
- National Meteorological Service.

We also expect International Support from:
- United Nations Office for Outer Space Affairs and the United Nations/European Space Agency Workshops on Basic Space Science;
- Scientific organizations;
- Foreign international programs to develop Science (European and others);
- the International Astronomical Union (UAI).

CONCLUSIONS

Astronomy can make an important contribution to general education in developing countries as well as developing some parallel interests in technical and technological studies and projects. It can also support the national space science program. The ongoing project TAURO, to install a mid-size automated and robotically controlled telescope in Uruguay, pursues these objectives.

Figure 1

DIRECCION DE RED DE ESTACIONES E INSTRUMENTAL
VIGILANCIA METEOROLOGICA MUNDIAL
SISTEMA MUNDIAL DE OBSERVACION
SUBSISTEMA DE SUPERFICIE
Bella Unión
Artigas
Rivera
Salto
Tacuarembó
Azucitrus
Puysandú
Paso de
los Toros
Melo
Young
Treinta y Tres
Durazno
Mercedes
Trinidad
Carmelo
Rocha
Libertad
El Colorado
Colonia
Melilla
Carrasco
Punta del Este
Central Prado
Maldonado
RED SINOPTICA BASICA REGIONAL, ESTACIONES TERRESTRES PRINCIPALES
ESTACIONES SUPLEMENTARIAS
ESTACIONES METEOROLOGICAS AERONAUTICAS
ESTACIONES CLIMATOLOGICAS PRINCIPALES
ESTACIONES CLIMATOLOGICAS ORDINARIAS
ESTACIONES METEOROLOGICAS AGRICOLAS
PARA FINES ESPECIALES
ESTACIONES RADIOMETRICAS PRINCIPALES
ESTACIONES RADIOMETRICAS ORDINARIAS

Figure 2

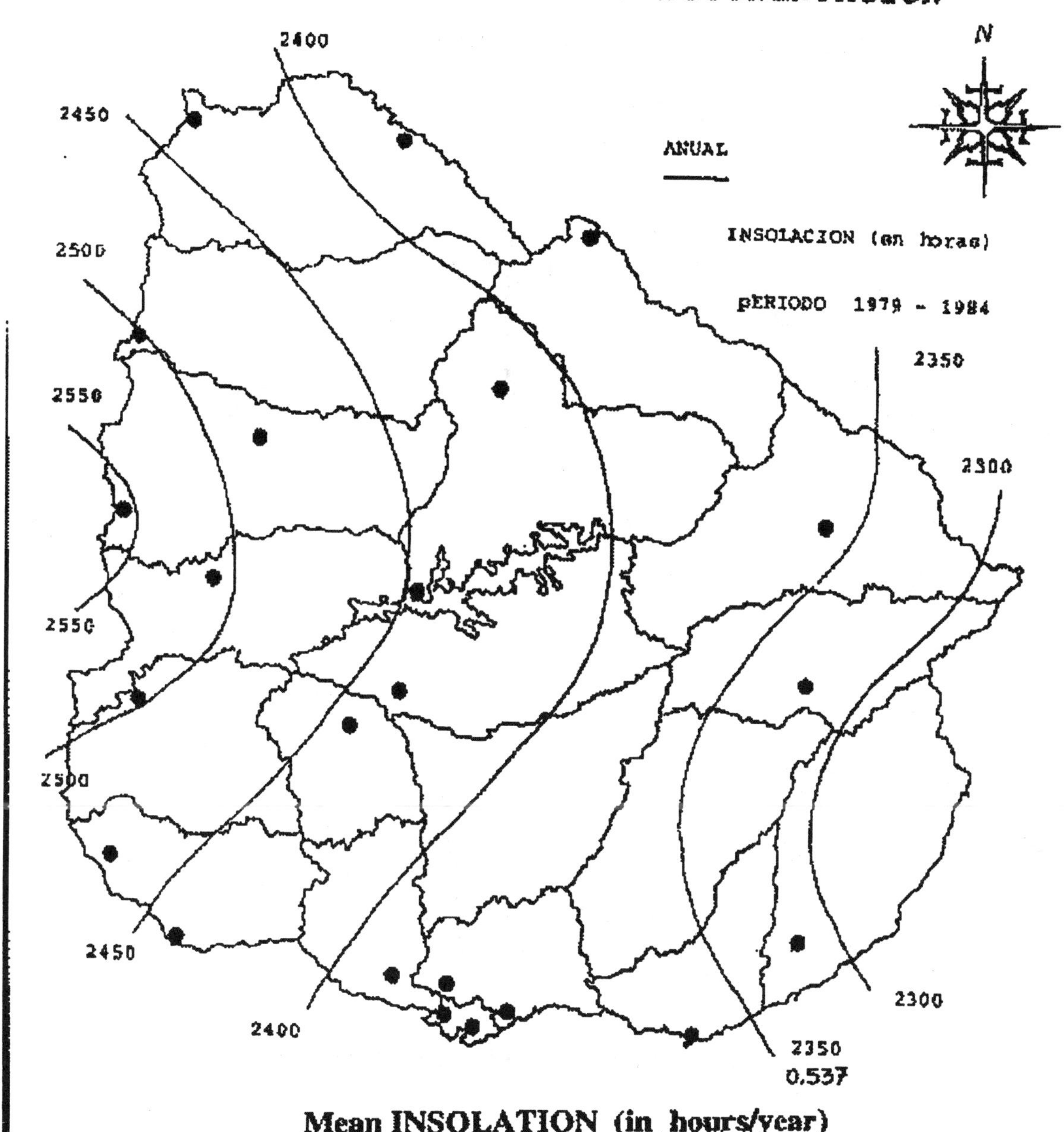

Mean INSOLATION (in hours/year)
note : 1 yr = 365x24 = 8760 hr.

HUMEDAD RELATIVA MEDIA ANUAL (%)

URUGUAY - 1961-1990

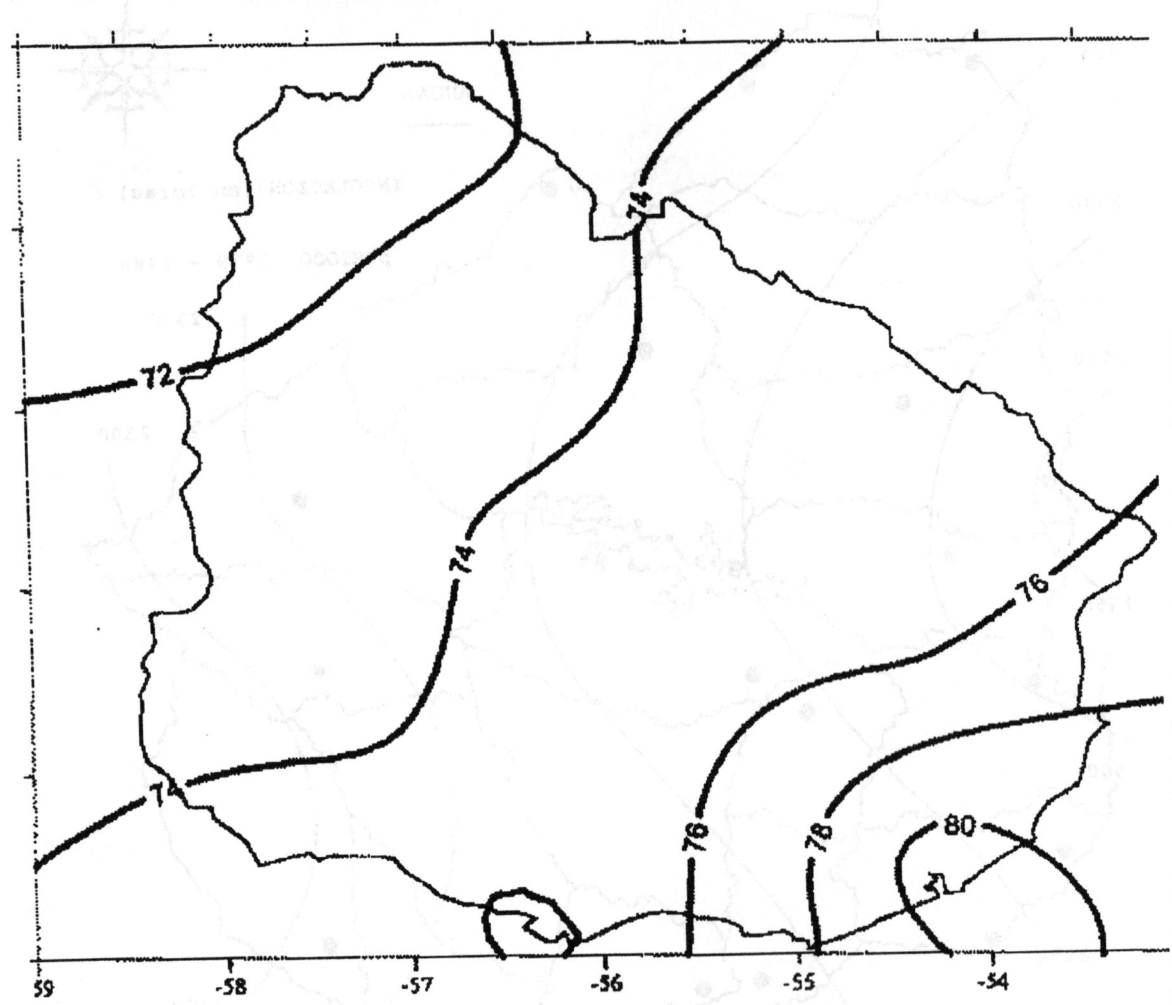

Annual Mean Relative Humidity (%)

Figure 4

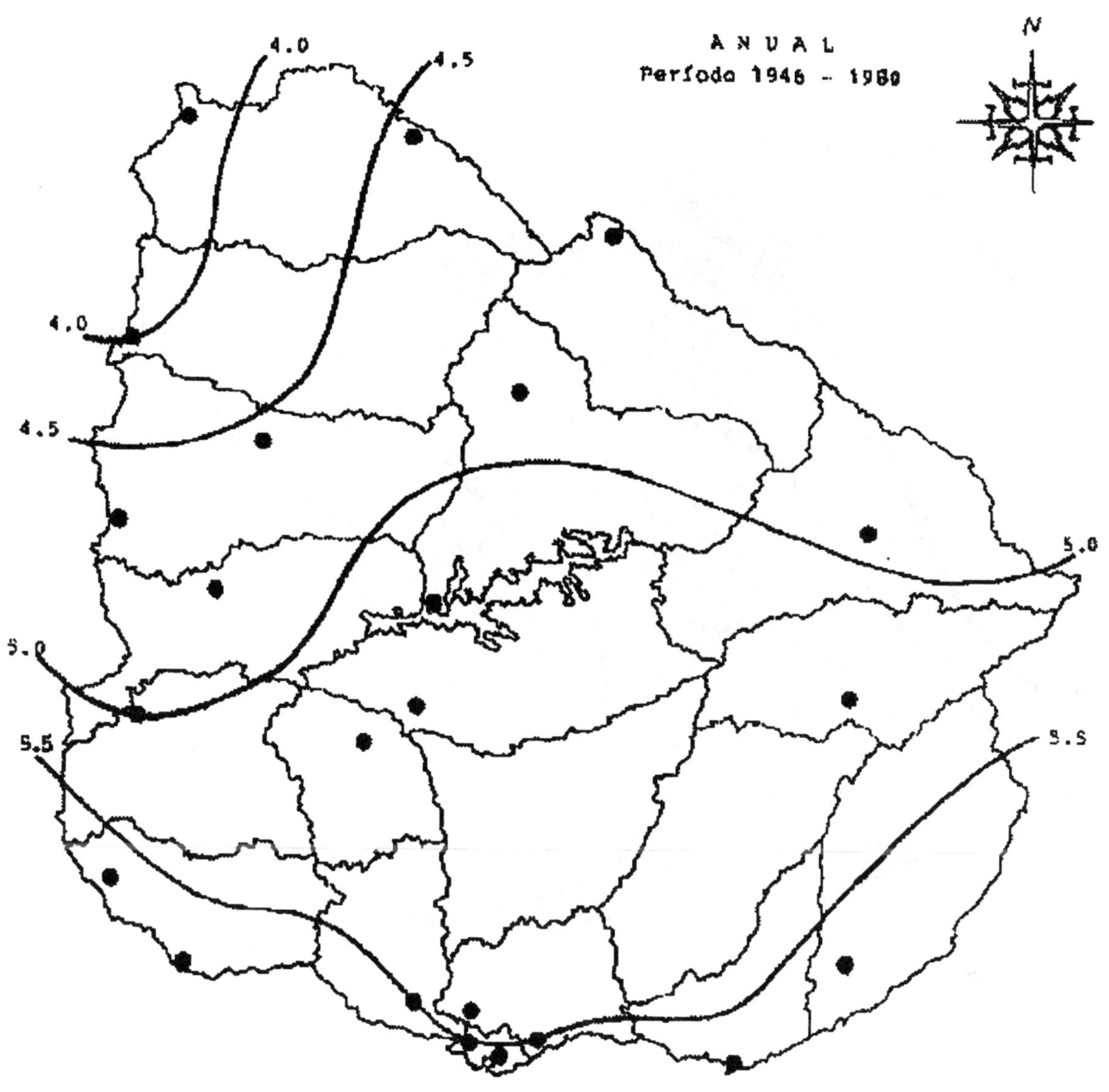

Mean CLOUDINESS (in eigthths of covered sky)

Figure 5

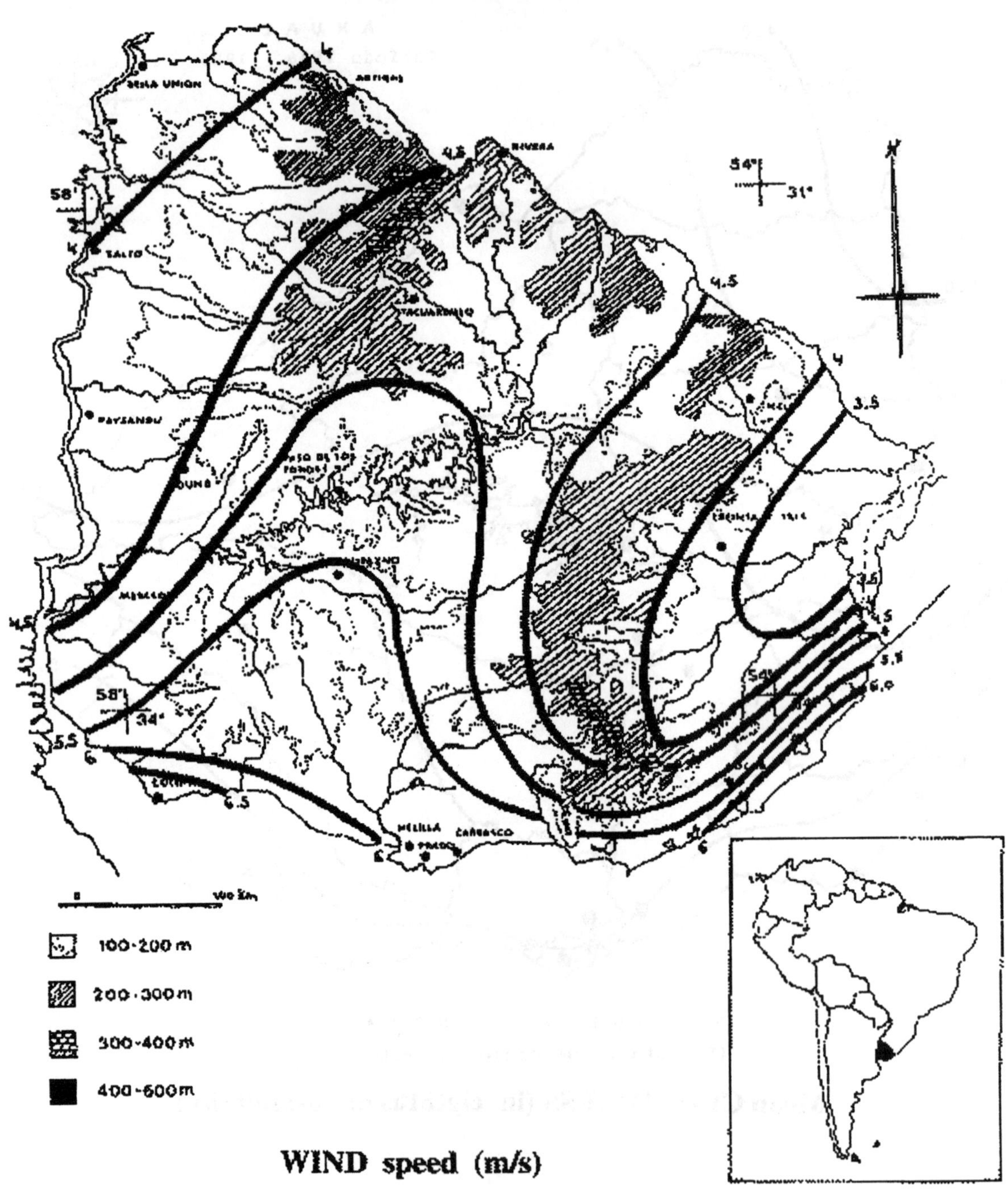

WIND speed (m/s)

Figure 6

09:40 a.m. 21/6/2000   WINTER SEASON
PRONÓSTICO, TARDE DEL MARTES
A
A
B
A
Cnn imagen de la presión
Yahoo imagen satelital.

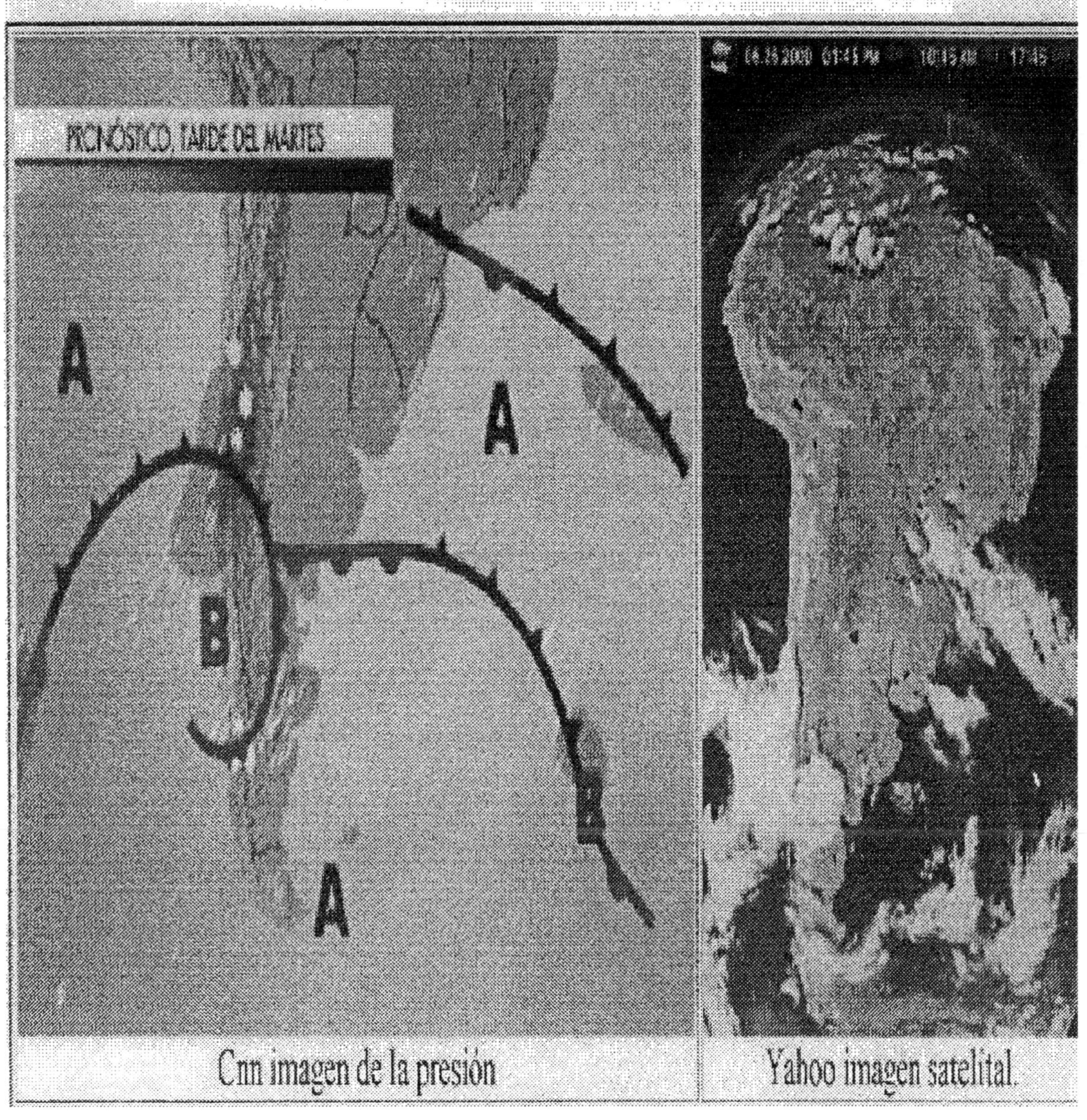

Figure 7: Map showing the location of Maldonado province

Figure 8: Landscape in Maldonado province

A NEAR-INFRARED CAMERA WITH A 512 x 512 PtSi CCD, COOLED BY TWO STIRLING CYCLE MACHINES, DEVELOPED FOR THE NISHI-HARIMA ASTRONOMICAL OBSERVATORY AND THE STATUS OF THE EDUCATIONAL ASTRONOMICAL OBSERVATORY PROJECT IN PERU[*]

José K. Ishitsuka I.[1], Takehiko Wada[2], Fumihiko Ieda[3], Noritaka Tokimasa[4], Takehiko Kuroda[4], Masaki Morimoto[4], Takeshi Miyaji[5], Toshihiro Omodaka[3], Munetaka Ueno[1] and Mutsumi Ishitsuka[6]

E-mail (JI): pepe@chianti.c.u-tokyo.ac.jp

ABSTRACT

The authors have developed a near-infrared CCD camera with a 512 x 512 PtSi detector. The camera dewar is refrigerated by two independent Stirling Cycle machines. The project was developed for the Nishi-Harima Astronomical Observatory of Japan, an observatory managed by Hyogo Prefecture and open to the general public. We began regular observations with the camera on April 24th, 2000. We present this camera as an alternative for carrying out astronomical observation using small telescopes and also as an alternative to robotic telescopes.

The Nishi-Harima Astronomical Observatory is also a supporter and a leader for donations to send a 60-cm class telescope to Peru for the projected Educational Astronomical Observatory. This paper will also present the progress with this project.

INTRODUCTION

The authors have developed a new infrared camera system that can be used to research radiation processes in circumstellar water vapor masers and for monitoring evolved stars. Observations in the infrared wavelength and simultaneous monitoring in the radio frequency of water vapor maser 22 GHz will produce important information about the poorly known radiation processes of water masers.

---

[*] *This paper was presented at the "Ninth United Nations/European Space Agency Workshop on Basic Space Science", held from 27 to 30 June 2000 in Toulouse, France, and does not necessarily reflect the views of the United Nations.*

[1] Department of Earth Sciences and Astronomy, University of Tokyo, Komaba, Tokyo 153-8902, JAPAN
[2] Institute of Space and Astronautical Science, Sagamihara, Kanagawa 229-8510, JAPAN
[3] Department of Physics, Kagoshima University, Kagoshima 890-0084, JAPAN
[4] Nishi-Harima Astronomical Observatory, Sayo, Hyogo 679-5313, JAPAN
[5] Nobeyama Radio Observatory of the National Astronomical Observatory of Japan, Minami-saku, Nagano 384-1305, JAPAN
[6] Instituto Geofisico del Perú, Observatorio de Ancón, PERU

The aim of the project is to develop the most simple camera system possible, one that is free of maintenance. It will be possible to install a similar system in the Southern Hemisphere, at the projected Educational Astronomical Observatory in Peru, if successful

THE CAMERA DEWAR

The cooling for the dewar that houses the 512 x 512 PtSi detector, which was tested successfully for astronomical observations by Ito et al. (1995), is equipped with two independent Stirling Cycle refrigerators. Their cooling capacity is 77 K, 1 watt and 50 K, 0 watts, respectively. As figure 1 shows, one of the refrigerators cools the cold plate where the chip is mounted. The second refrigerator cools the thermal radiation stage.

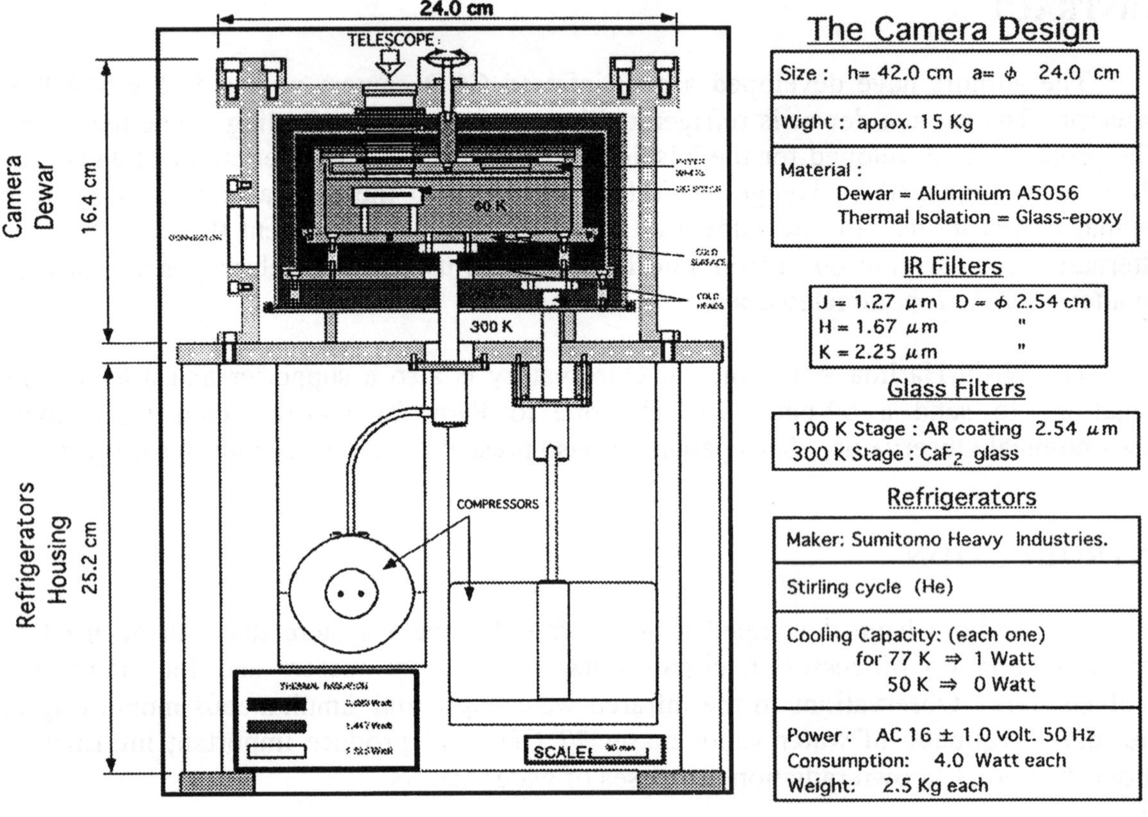

Figure 1: The camera system

The use of these kinds of refrigerators will improve observations with less maintenance. This means that there is no need to use refrigerant; the camera cools down using the refrigerators. This is a good alternative for robotic telescopes or remotely driven systems. Our system is designed so that it can be connected and driven through the Ethernet, as shown in Figure 2.

This project is supported by the Nishi-Harima Astronomical Observatory, which is managed by the Hyogo Prefecture of Japan, Kagoshima University, the University of Tokyo, the Nobeyama Radio Observatory of the National Astronomical Observatory of Japan, and the Institute of Space and Astronautical Science.

The Near Infrared CCD Camera NIHCOS - (Nishi-Harima Cosmos)

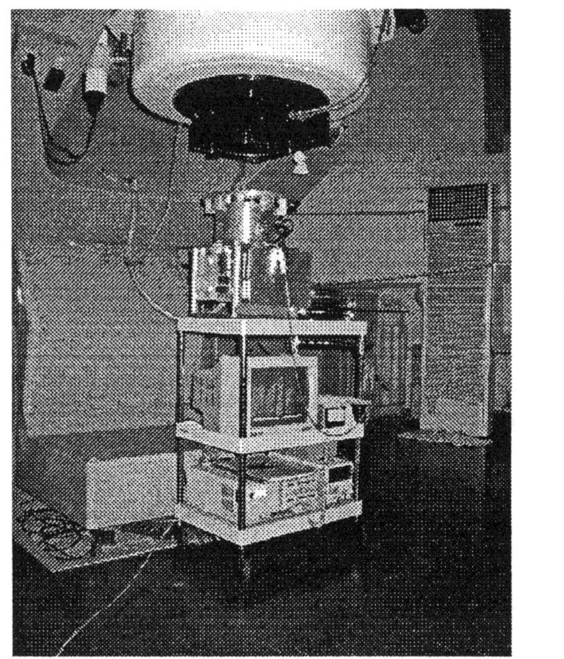

NIHCOS + 512 x 512 PtSi @ Nishi-Harima 60 cm telescope

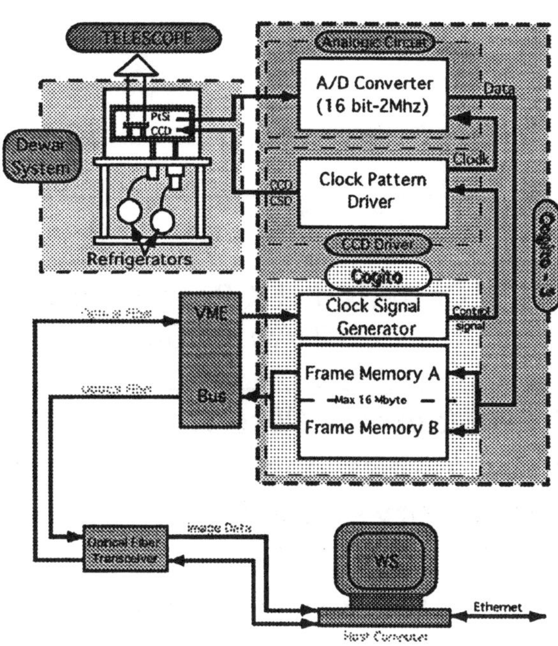

The Camera System

Figure 2.: A photo (left) of the infrared camera to be installed in the telescope, and a diagram (right) of the whole system.

We began regular infrared observations of the semi-regular variable star R Crateris on April 24th, 2000, a star that has been monitored with the Japanese VLBI Network (J-Net) by Ishitsuka et al. (2000), where proper motion of water vapor masers was detected. Our infrared observations aim to determine the light curve of R Crateris, at the three bands J, H and K, and also to find the correlation with maser emission intensity observed at radio frequencies. Preliminary results are presented in Figure 3.

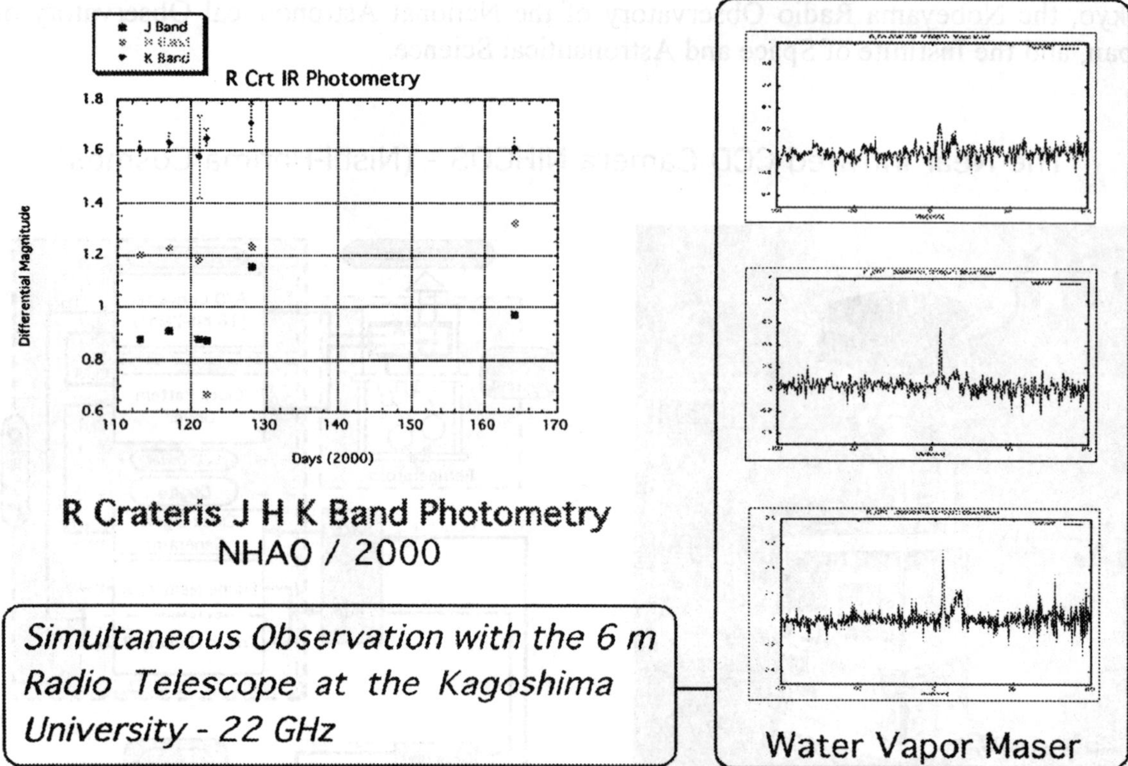

Figure 3: The first six photometric measurements at the J, H and K bands.

Since the main function of the Nishi-Harima Observatory is to promote astronomical knowledge among the general public, some infrared images, for instance of planets, galaxies, globular clusters and binary stars, have been employed for regular publications of the Observatory. In addition, a simultaneous observation of the binary star RS Vul was carried out at the V band, using a CCD of the amateur astronomer Yoshifumi Waki, who is a member of the Association of Amateur Astronomers, directed by the Nishi-Harima Observatory. Results are shown in Figure 4.

STATUS OF THE EDUCATIONAL ASTRONOMY OBSERVATORY IN PERU

Mutsumi Ishitsuka, Director of the Observatorio de Ancón of the Instituto Geofisico del Perú, is in charge of developing and constructing the Educational Astronomical Observatory. After searching for an ideal location for the observatory, which should be relatively near to the capital Lima and should have good observational conditions, it was tentatively decided to place the observatory 270 km south of the capital, a few kilometers from the Pan-American Highway. Some details can be seen in figure 5.

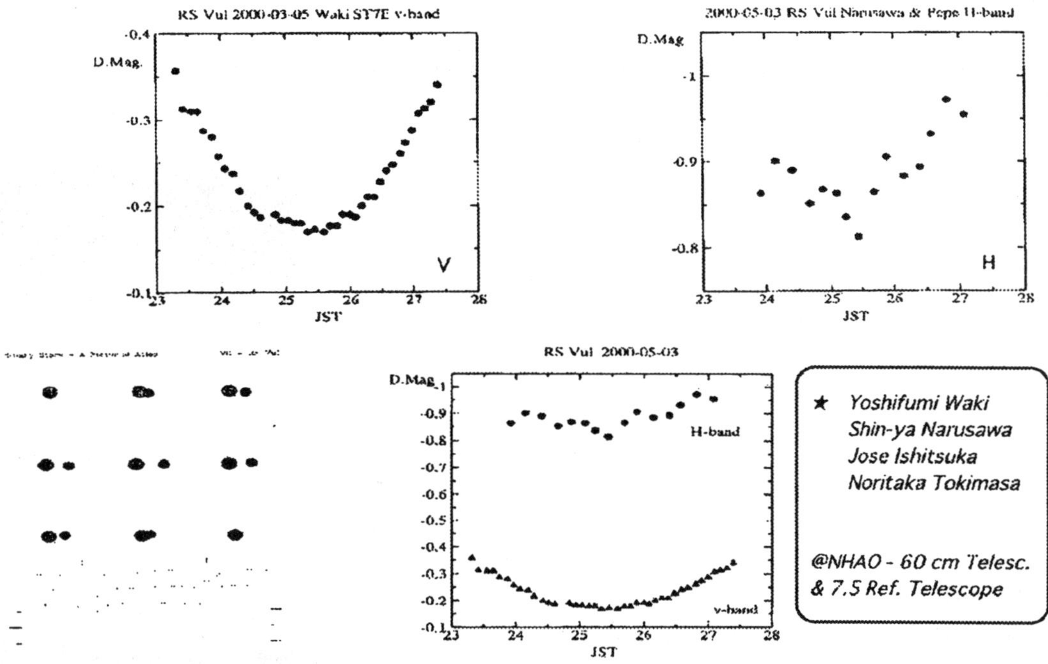

Figure 4.: Rs Vul photometry test, differential magnitudes light curve at H and K bands.

The Observatory will have a capacity of 25 visitors, an auditorium, a library and Internet facilities. It will be equipped with two 15-cm aperture refractor telescopes, both of which are already used for regular observations and were donated by the Japanese government in 1994. In addition, a 60-cm class telescope is planned to be installed. The Director of the Nishi-Harima Observatory, Mr. Takehiko Kuroda, is leading a search for donations from Japan to buy the telescope and to donate it to the projected Observatory.

Dr. Masatoshi Kitamura, Professor Emeritus at the National Astronomical Observatory of Japan and a supporter of the project, personally visited the prospective sites and gave important suggestions. He is also actively participating in the process to install a new planetarium in Peru.

CONCLUSIONS

The authors developed a new near-infrared camera, cooled by two independent Stirling Cycle refrigerators and using a 512 x 512 PtSi CCD chip as a detector. The camera works very well and is an alternative to small telescopes. It is also a prototype camera that extends the horizon for remote-controlled ground base infrared telescopes.

The new project in Peru to construct an Educational Astronomical Observatory is becoming a reality, thanks to efforts of the Governments of Japan and Peru. Furthermore, the unconditional cooperation of Dr. M. Kitamura and Mr. T. Kuroda has made a valuable contribution to enhancing astronomical education in Peru.

☆ *An Educational Astronomical Observatory Project - Perú*

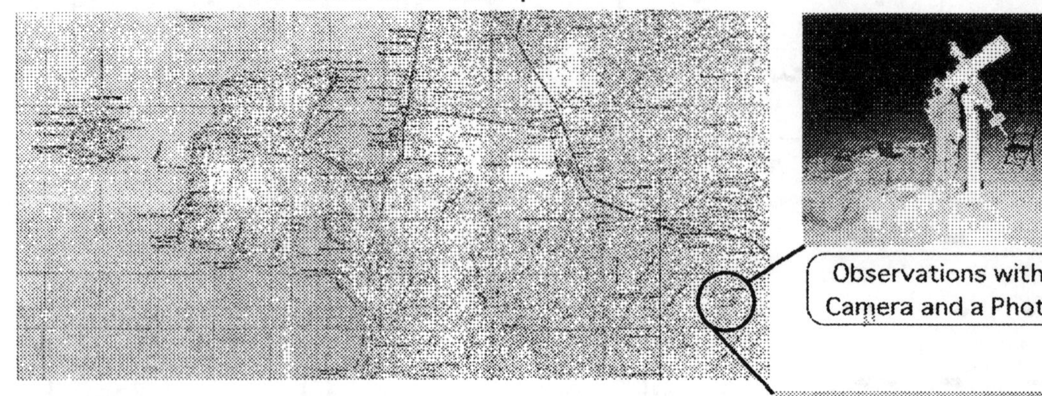

Figure 5.: The site of the projected Educational Astronomical Observatory. Photos are of test observations using the 15-cm telescope donated by the Government of Japan.

REFERENCES

● Ishitsuka, J., Imai, H., Omodaka, T., Ueno, M., Kameya, O., Sasao, T., Morimoto, M., Miyaji, T., Nakajima, J., Watanabe, T. and J-Net Members, 2000, VLBI Monitoring Observations of the Semi-regular Variable Star R Crateris, paper in preparation.
● Ito, M., Kasaba, Y., Ueno, M. and Sato, S., 1995, ASP, 107, 691.

DISASTER REDUCTION IN THE TWENTY-FIRST CENTURY*

Francesco Pisano
Secretariat for the United Nations International Strategy for Disaster Reduction (ISDR)

Excellencies, Distinguished Delegates and Representatives, Colleagues, Ladies and Gentlemen:

I am very pleased to have been invited to participate in this important meeting on behalf of the United Nations International Strategy for Disaster Reduction (ISDR). Let me first of all congratulate the Government of Chile (MOFA) and the beautiful town of La Serena and its University for hosting this event. I also wish to thank the United Nations Office for Outer Space Affairs (OOSA) and the European Space Agency for the invitation extended to us.

I see in the audience friends and colleagues with whom we have been working, both as the International Decade for Natural Disaster Reduction (IDNDR) and now as ISDR, to advance the use of satellite techniques in disaster reduction. The growing interest for this field of activity is a testimony of the potential role of satellite applications in the reduction of the impact that natural hazards have on society and the economy.

Today I am requested to present the latest international developments in the field of disaster reduction by explaining the nature and scope of the new ISDR Programme of the United Nations and mention some of the issues recognised by the international community as accepted priorities.

I do not need to underline that we meet here today against the background of the increasing incidence of disasters, the most recent manifestation of which are the widespread floods which are occurring in many parts of Europe. Chile itself is prone to a wide range of hazards and has a long record of experiences of their painful consequences. It is in fact becoming more and more evident that the increase in the frequency and intensity of earthquakes, floods and other disasters, suggests that our planet is faced with a growing threat.

This is not empty rhetoric, since an analysis of the trend in major natural disasters occurring between 1950 and 1999 indicates an exponential growth in the occurrence of such disasters, particularly during the past decade. In 1999 alone there were more than 700 large-scale disasters with losses in excess of 100 billion USD. We at ISDR sincerely believe that the time has come to recognise the growing danger posed by this situation and the need to treat disaster reduction as a central issue on the international agenda.

---

* *Keynote Address at the United Nations/Chile/European Space Agency Workshop on the "Use of Space Technology in Disaster Management", held from 13 to 17 November 2000 in La Serena, Chile.*

As most of you know, it was this recognition that led in part to the launch of the International Decade for Natural Disaster Reduction (IDNDR) during the period 1990-1999. The various initiatives promoted during the Decade have served to create increased awareness of the importance of disasters, whether natural, environmental or technological. Many organisations and numerous countries, and Chile is surely among them, joined efforts during the IDNDR to bring the issue of disaster reduction to the fore of the international debate.

What is most significant about this process is the recognition that although many of these phenomena are referred to as natural disasters, human behaviour is the major cause of our increasing vulnerability to natural hazards. Climate change, deforestation and changing demographic trends emphasise the role of the human factor in creating a predisposition to the increased severity and frequency of disasters. In other words, human activities contribute to creating vulnerability, increasing the risk we face that natural phenomena turn into actual disasters.

It is this recognition that led to a significant conceptual shift from the traditional preoccupation with disaster response to a new culture of disaster reduction, which is reflected in the United Nations programme named the International Strategy for Disaster Reduction.

Based on the wealth of knowledge and experience accumulated during the International Decade, our new vision is to proceed from the protection against hazards to the management of risk. The ISDR programme is a pro-active mechanism conceived to facilitate the design, implementation and evaluation of effective disaster prevention and vulnerability reduction measures with a view to reducing the impact of natural and other hazards on society and the economy.

The adoption of this new vision by the Member States of the United Nations has brought about a new thrust directed to elicit *action* and financial and political commitment from both national governments and international organisations. This is why the ISDR is mainly an inter-agency endeavour.

I should explain that the International Strategy for Disaster Reduction is a global strategy with unusually broad membership. Institutionally, it has two arms.

The first arm is the Inter-Agency Task Force, which represents a rare three-way coalition between regional entities, UN agencies and representatives of civil society. The Task Force carries forward strategies and policies for the reduction of natural hazards; it identifies gaps in disaster-reduction policies and programmes; and it ensures that there is an appropriate division of labour among agencies and organisations so that resources are not wasted on overlapping programmes.

The second arm is the Inter-Agency Secretariat, which is based in Geneva and serves as the focal point within the United Nations for strategies and programmes for natural-disaster reduction. It is a multi-disciplinary support team that supports the forum in which international policy can be made. It is also the institutional platform from which programmes can be launched. It doesn't implement programmes, but allows others to implement programmes more effectively.

Following the recent Inter-Agency Task Force meeting the Secretariat is formulating a plan of action based on the identification of a number of strategic initiatives to be undertaken at the global, subregional/regional and national level, including the provision of support to the national ISDR platforms. These initiatives are designed to complement, and not to duplicate, the work of the various international organisations active in disaster reduction through their respective mandates.

I mention these developments in order to provide the overall context for my more specific comments on the issue of space applications for disaster reduction, which is the subject of this meeting.

The field of satellite technology has several points in common with the ISDR philosophy. Both presume a high degree of international cooperation in dealing with multi-agency programmes leading to integrated solutions to a problem that has a growing global dimension. This morning Ambassador Winter mentioned the importance of international cooperation. Most disasters are in fact transboundary, both because of the variety of factors that determine their pattern, and because of the fact that their consequences spread across national borders. The programmes being put forward by Committee on Earth Observation Satellites (CEOS), OOSA, the European Space Agencies and other entities are a response to the need to take into account the international dimension of disaster reduction.

In yet another example, developing satellite based solutions to obviate to the vulnerability of communities world-wide to natural hazards responds to one central feature of our work at ISDR: using technology applications to help societies to become resilient to natural and other hazards. Vulnerability is created at the local level. We need to proceed from international strategies to regional and local solutions, if we are to be successful in the long run.

While international cooperation is a fundamental component of any programme for the wide dissemination of satellite data, we must also think of fostering national and regional research programmes, encourage the development of modelling tools for the generation of indicators and risk maps, promote the use of early warning systems and educate communities to react properly to warnings.

We must also anticipate the compound effects of natural and other technological and environmental hazards: in future the nature and the size of disasters may force us to revisit our notion of vulnerability and risk management and reconsider the relationship between prevention and emergency relief as we conceive it today.

In the words of the Secretary-General of the United Nations, disaster prevention must become an integral part of development planning at all levels, informing our future endeavours in fields as varied as land use planning, environmental management and the protection of natural resources.

Seen from this angle, it clear that much remains to be done. Yet I think we can say that we have come a long way in evaluating the potential of satellite techniques for disaster reduction, assessing user requirements, and exploring the feasibility of integrated solutions.

I should mention in this context the work done by the CEOS Disaster Management Support Project, which evolved into a group of international experts that has become a reference in this domain. I should also mention our partnership with OOSA, which carries forward a programme focused on disaster reduction. And of course the European Space Agency and several national space agencies, which are conducting a promising work that will certainly be illustrated during this meeting.

The ISDR has placed increased emphasis on satellite applications in recognition of the significant impact that these innovative solutions can have in terms of risk management for vulnerable communities in both developed and developing countries. Of course, we cannot stress enough the importance of using these applications to the benefit of developing countries, where the socio-economic impact of disasters threatens hard-earned development assets. More than one speaker underlined this aspect this morning.

In reviewing the programme of this meeting I was impressed not only by the range of issues that will be discussed, but also by the participation of experts from several countries and organisations which should contribute to a cross fertilisation of ideas on this important subject. Equally important is the (significant) emphasis placed by the meeting on potentially innovative processes for the use of satellite and earth observation data in disaster management. This clearly reflects the possibility that a difference can be made and that the application of research, science and technology can play an important role in promoting the objectives of disaster reduction.

Given the importance of this issue, we at ISDR look forward to working with our partners in this field to provide an international framework for the design and application of disaster reduction measures.

In conclusion, let me say that the ISDR Secretariat is committed to working with countries, regional institutions and all other stakeholders to advance the common goal of facing the global challenge of preserving our societies from the disruptive consequences of disasters. This morning Professor Pozo, Mr. Fuentealba and others pointed us to the importance of the human aspects of disaster reduction; to the fact that we have to work without loosing sight of our ultimate goal: to reduce human suffering and achieve equitable, sustainable human development. This is in short one of the aims of the United Nations. I think we can all agree that we collectively owe this to future generations.

Thank you for your attention.

THE USE OF EARTH OBSERVING SATELLITES FOR HAZARD SUPPORT*

Helen Wood
National Oceanic and Atmospheric Administration (NOAA)

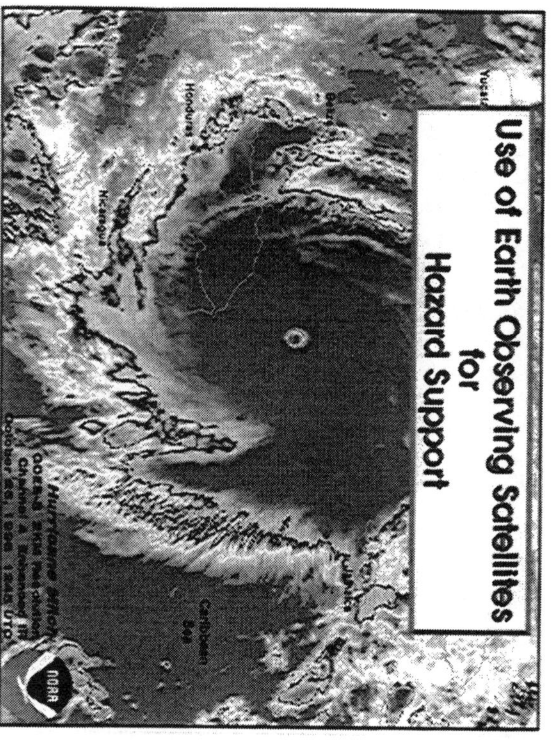

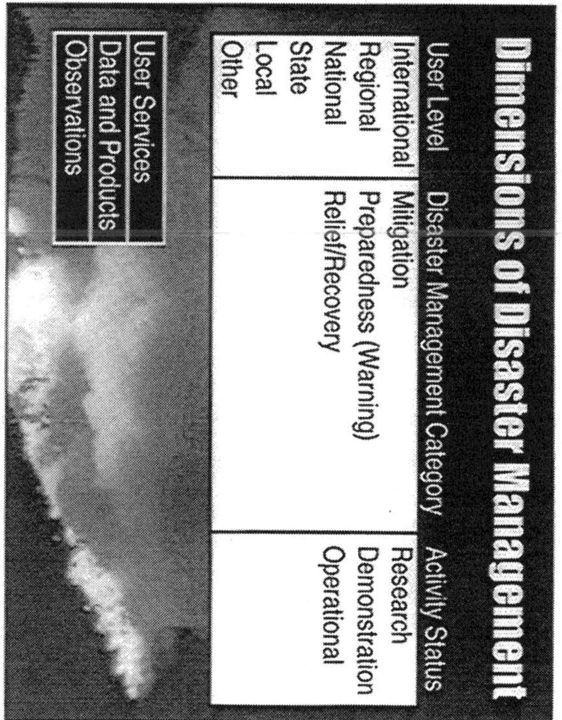

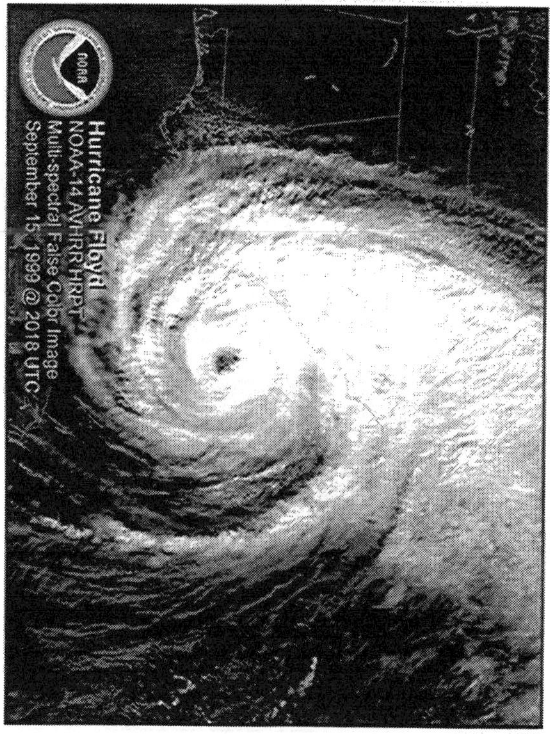

** A presentation of the Committee on Earth Observation Satellites (CEOS) Disaster Management Support Group at the United Nations/Chile/European Space Agency Workshop on the "Use of Space Technology in Disaster Management", held from 13 to 17 November 2000 in La Serena, Chile.*

IKONOS Image of Tornado Damage, Fort Worth, TX

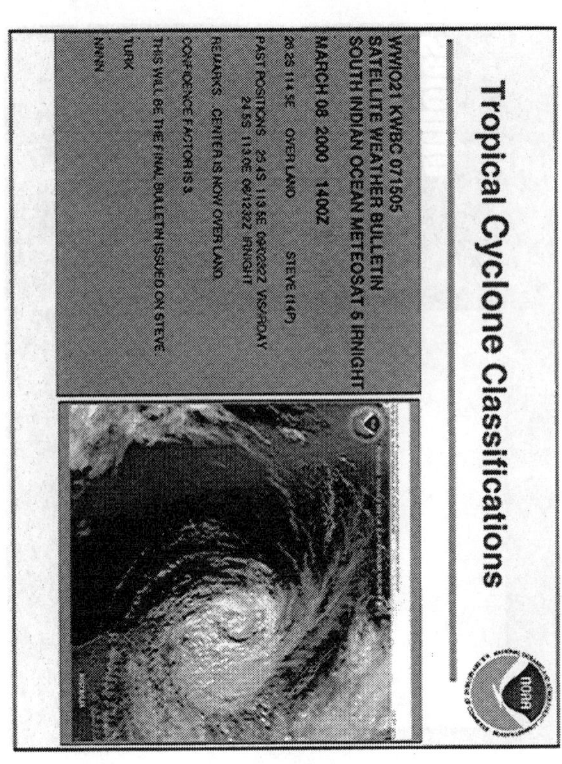

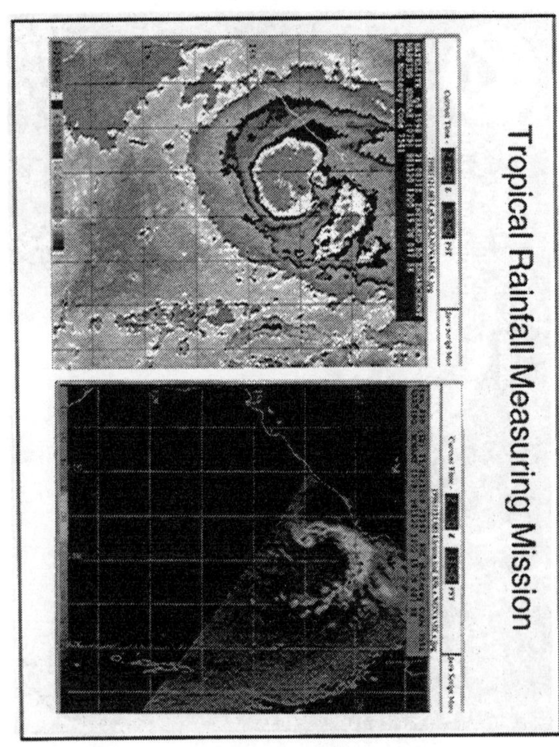

Tropical Rainfall Measuring Mission

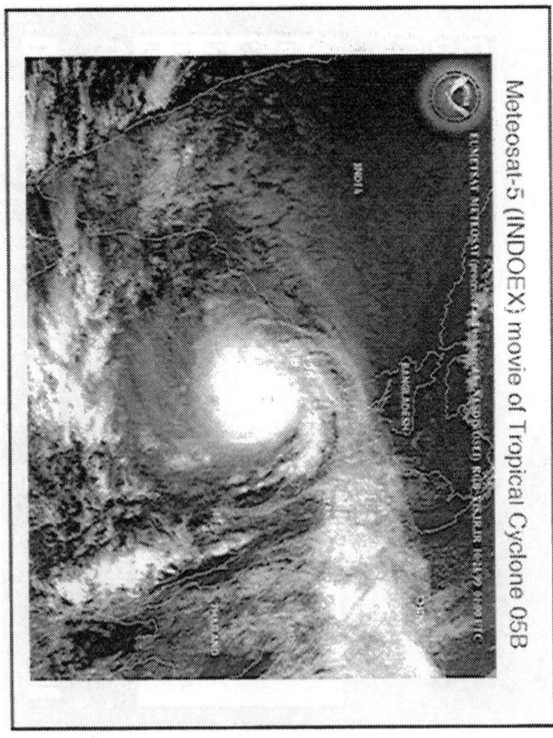

Meteosat-5 (INDOEX) movie of Tropical Cyclone 05B

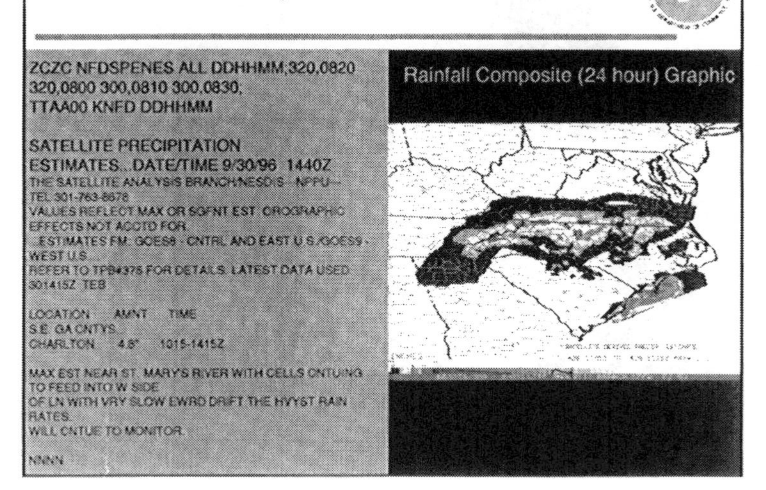

Nighttime Lights Differences Provide
Assessment of Storm's Impact on Power Grid
Winter
Ice Storm
Aftermath
Jan 18, 1998
Observed
Changes In
Nighttime
Lights
Source:
National
Geophysical
Data Center
Clouds .... Blue
Power Outage .... Red
Power Present
or Restored .... Yellow
Ohio
Montreal
Maine
New York
Vermont
NH
Lake Ontario
Boston

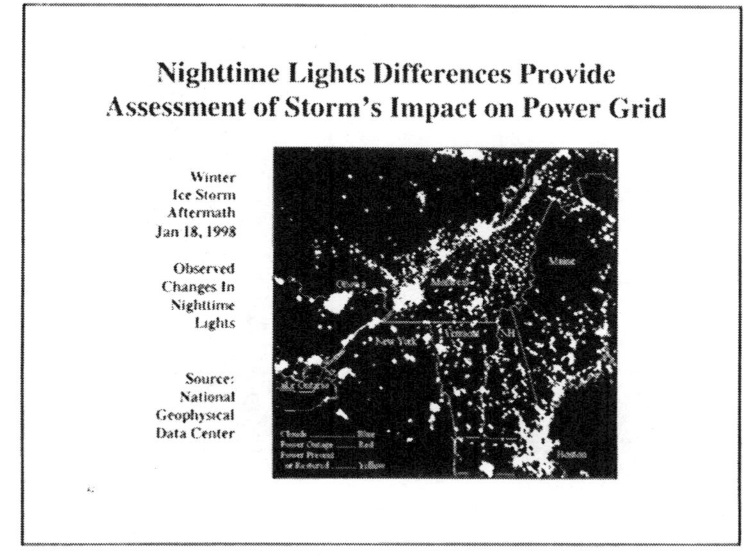

Precipitation Estimates
NOAA/NESDIS RAINFALL AUTOESTIMATOR PRODUCT

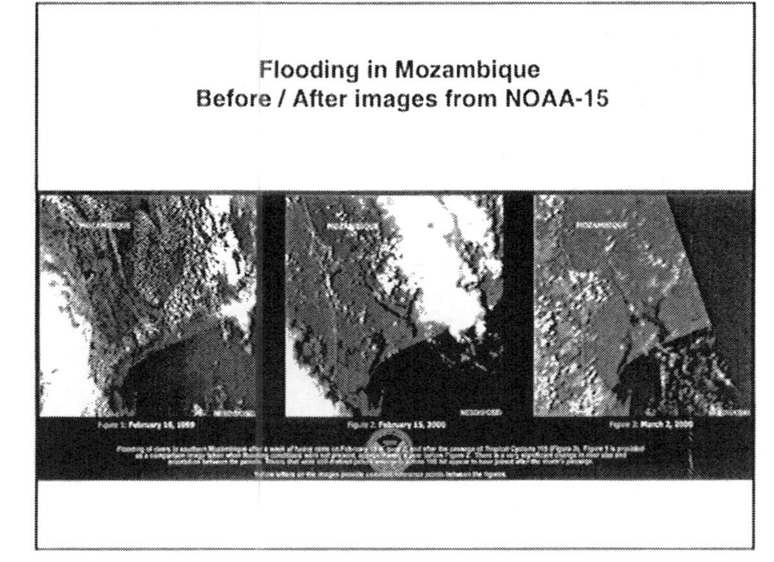

Precipitation Estimates
NOAA
Rainfall Composite (24 hour) Graphic
ZCZC NFDSPENES ALL DDHHMM;320,0820
320,0800 300,0810 300,0830;
TTAA00 KNFD DDHHMM

SATELLITE PRECIPITATION
ESTIMATES...DATE/TIME 9/30/96 1440Z
THE SATELLITE ANALYSIS BRANCH/NESDIS - NPPU--
TEL 301-763-8678
VALUES REFLECT MAX OR SGFNT EST OROGRAPHIC
EFFECTS NOT ACCTD FOR
ESTIMATES FM GOES8 - CNTRL AND EAST U S /GOES9 -
WEST U.S.
REFER TO TPB#375 FOR DETAILS. LATEST DATA USED.
301415Z TEB

LOCATION   AMNT   TIME
S.E. GA CNTYS.
CHARLTON   4.8"   1015-1415Z

MAX EST NEAR ST. MARYS RIVER WITH CELLS ONTUING
TO FEED INTO W SIDE
OF LN WITH VRY SLOW EWRD DRIFT THE HVYST RAIN
RATES.
WILL CNTUE TO MONITOR.

NNNN

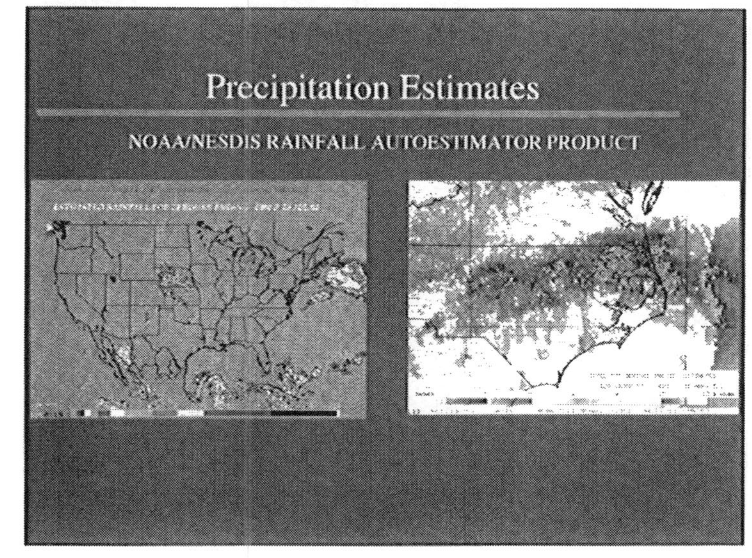

Flooding in Mozambique
Before / After images from NOAA-15
Figure 1: February 16, 1999
Figure 2: February 15, 2000
Figure 3: March 2, 2000

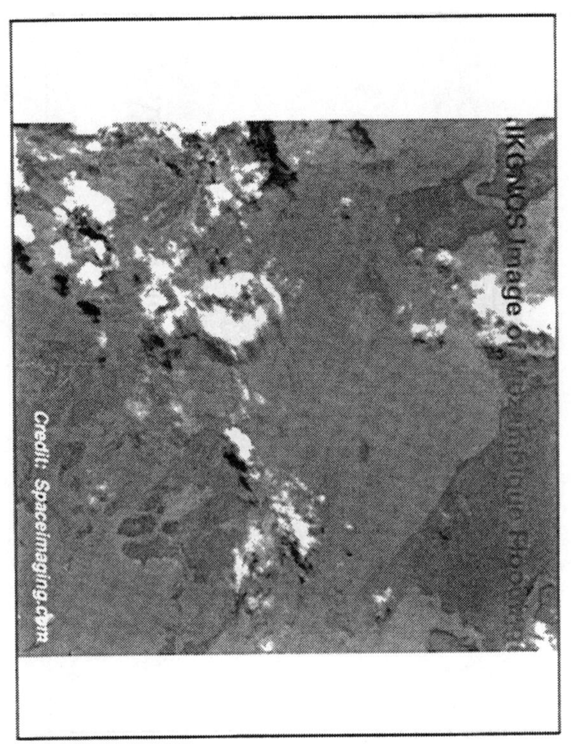

IKONOS Image of Mozambique Flooding
Credit: Spaceimaging.com

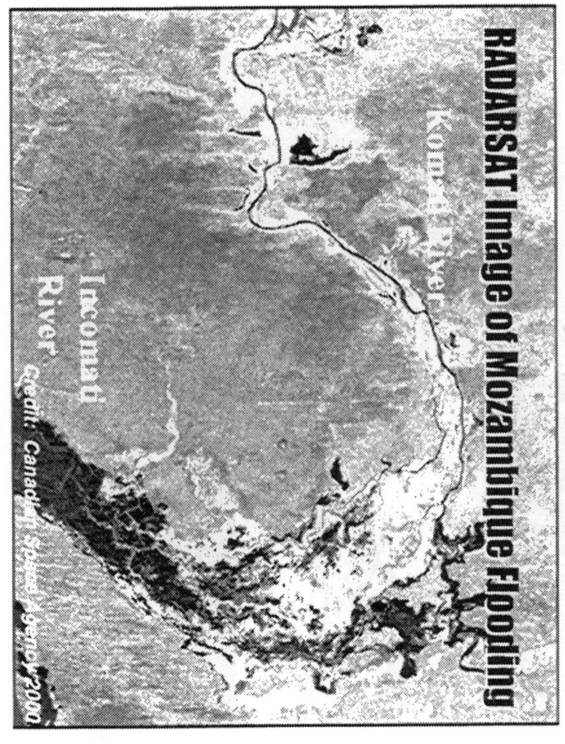

RADARSAT Image of Mozambique Flooding
Komati River
Incomati River
Credit: Canadian Space Agency, 2000

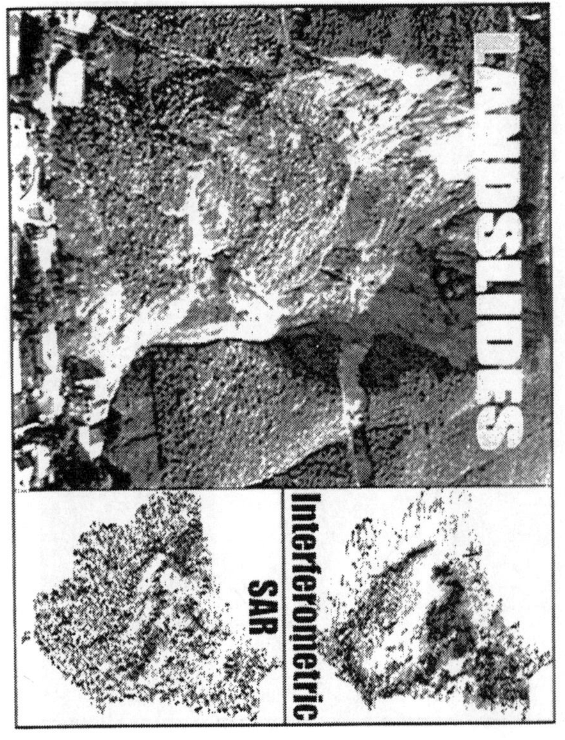

LANDSLIDES
Interferometric SAR

Earthquakes
GPS: Monitoring Southern California
ERS-1 SAR: Landers Earthquake
USGS
MO

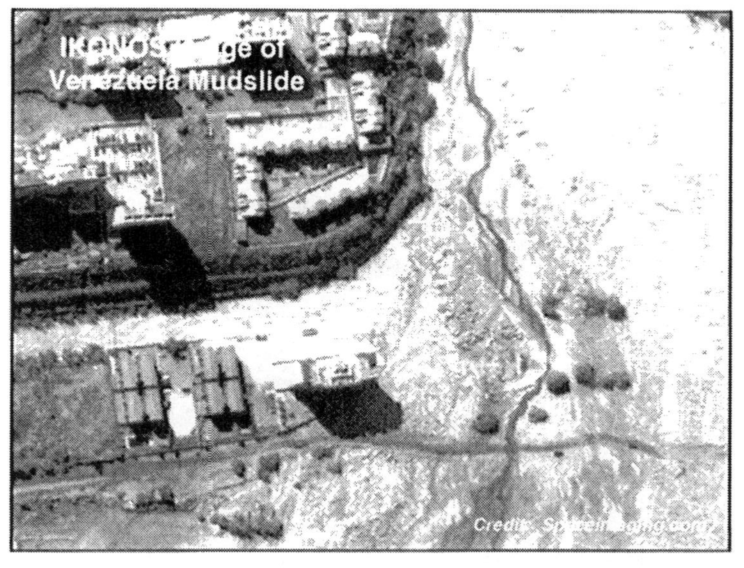

IKONOS Image of
Venezuela Mudslide
Credit: Spaceimaging.com

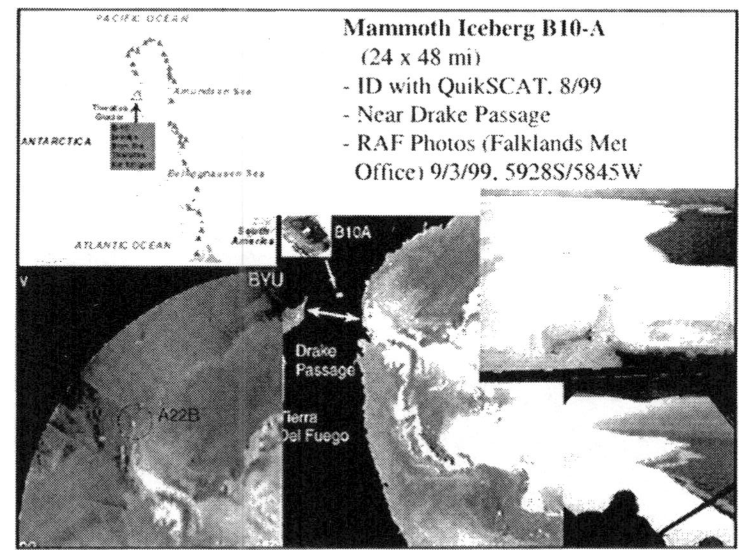

PACIFIC OCEAN
Amundsen Sea
ANTARCTICA
Bellingshausen Sea
ATLANTIC OCEAN
South America
Mammoth Iceberg B10-A
(24 x 48 mi)
- ID with QuikSCAT. 8/99
- Near Drake Passage
- RAF Photos (Falklands Met
Office) 9/3/99. 5928S/5845W
B10A
BYU
A22B
Drake Passage
Tierra Del Fuego

Sea Ice Hazards
Credit: NOAA Office of NOAA Corps Operations 1977

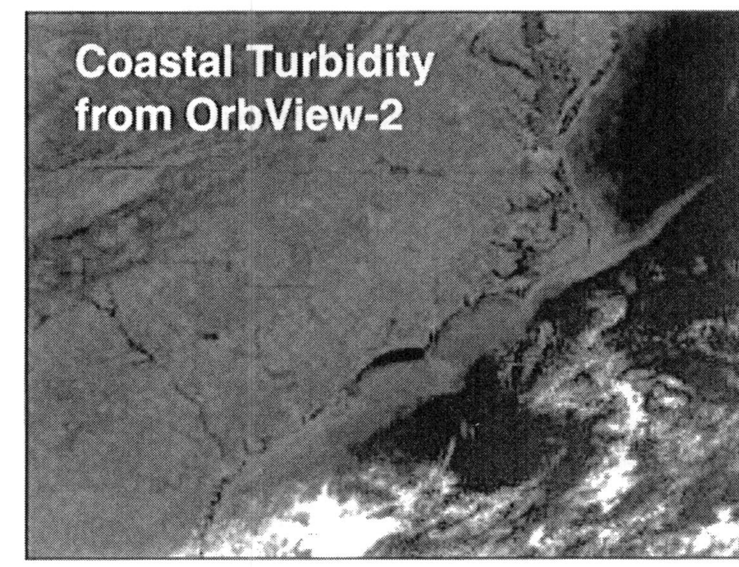

Coastal Turbidity
from OrbView-2

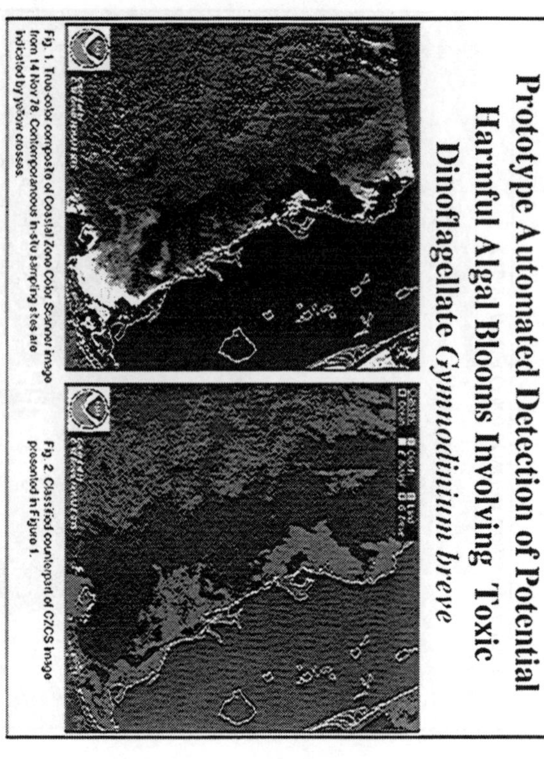

Prototype Automated Detection of Potential Harmful Algal Blooms Involving Toxic Dinoflagellate Gymnodinium breve
Fig. 1. True-color composite of Coastal Zone Color Scanner image from 14 Nov 78. Contemporaneous in-situ sampling sites are indicated by yellow crosses.
Fig. 2. Classified counterpart of CZCS image presented in Figure 1.

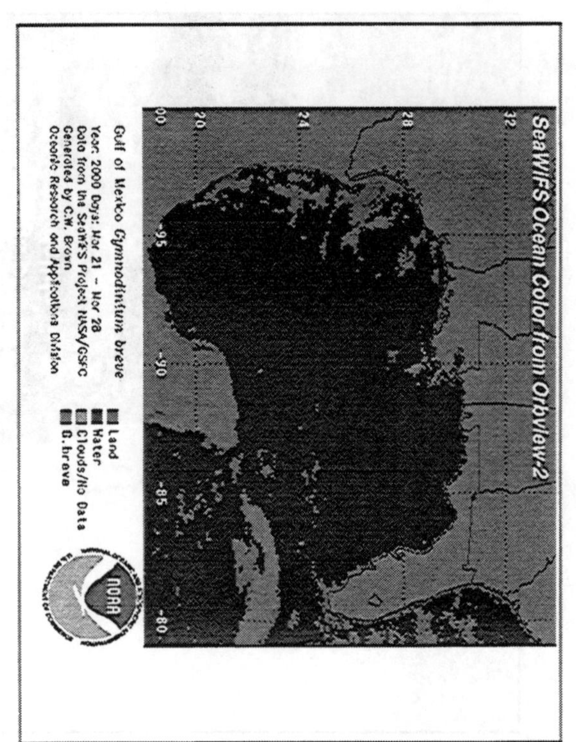

SeaWiFS Ocean Color from Orbview-2
Gulf of Mexico Gymnodinium breve
Year: 2000 Days: Mar 21 - Mar 28
Data from the SeaWiFS Project NASA/GSFC
Generated by C.W. Brown
Oceanic Research and Applications Division
Land
Water
Clouds/No Data
G. breve

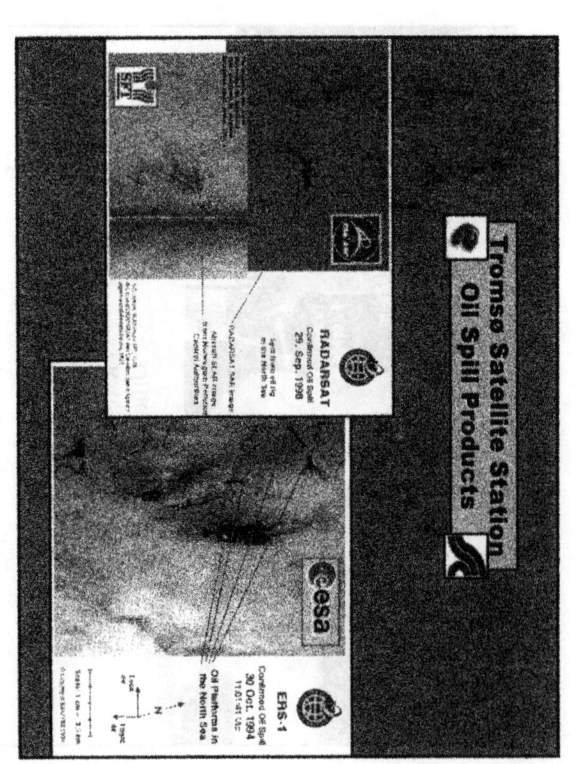

Tromsø Satellite Station
Oil Spill Products
RADARSAT
Confirmed Oil Spill
29. Sep. 1998
ERS-1
Confirmed Oil Spill
30 Oct. 1994
11:01:41 UTC
Oil Platforms in the North Sea

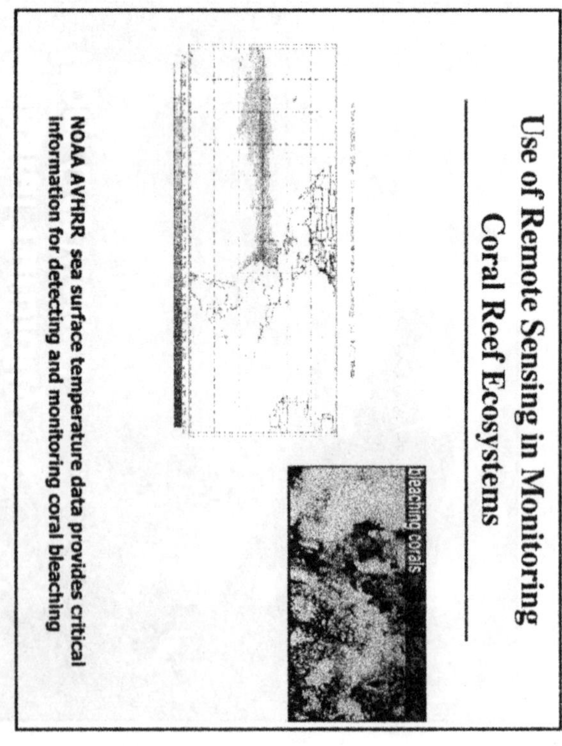

Use of Remote Sensing in Monitoring Coral Reef Ecosystems
NOAA AVHRR sea surface temperature data provides critical information for detecting and monitoring coral bleaching
bleaching corals

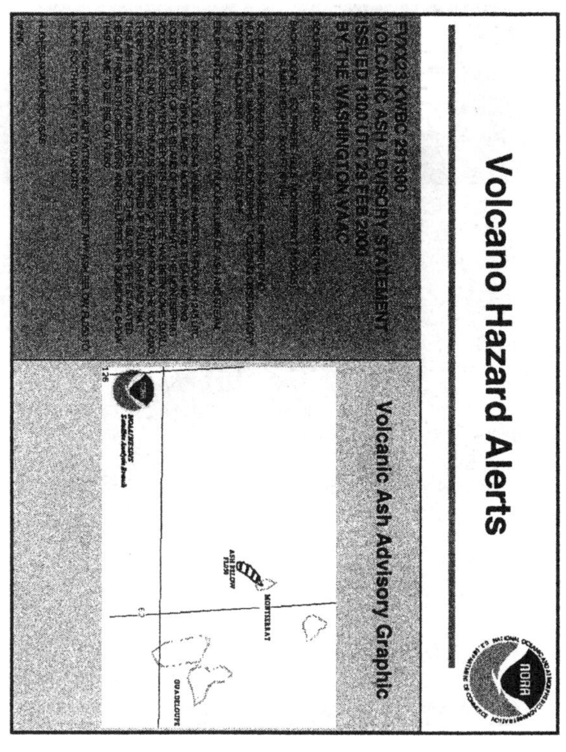

Volcano Hazard Alerts
Volcanic Ash Advisory Graphic
FVXX23 KWBC 291300
VOLCANIC ASH ADVISORY STATEMENT
ISSUED 1300 UTC 29 FEB 2001
BY THE WASHINGTON VAAC
NOAA/NESDIS Satellite Analysis Branch
ASH BELOW FL55
MONTSERRAT
GUADELOUPE

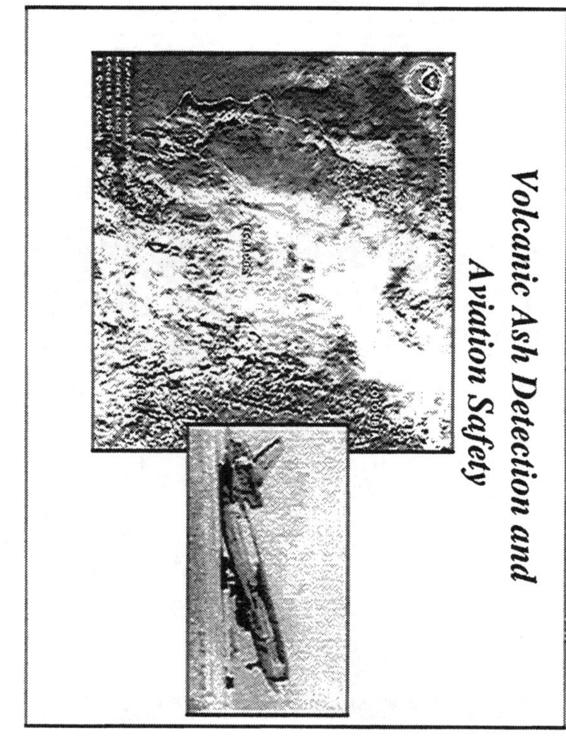

Volcanic Ash Detection and
Aviation Safety

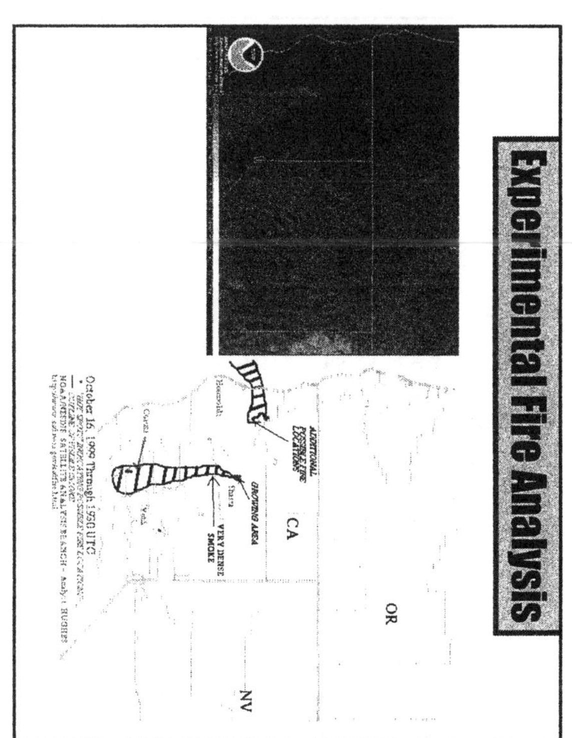

Experimental Fire Analysis
October 16, 1999 Through 1950 UTC
ADDITIONAL
DETRACT
VERY DENSE
SMOKE
BURNING AREA
Cortez
Yreka
Sierra
CA
OR
NV

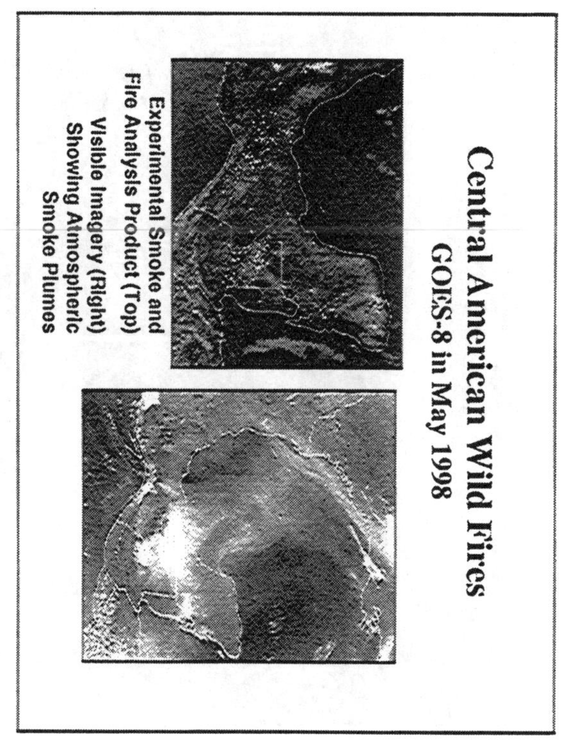

Central American Wild Fires
GOES-8 in May 1998
Experimental Smoke and
Fire Analysis Product (Top)
Visible Imagery (Right)
Showing Atmospheric
Smoke Plumes

SPOT 4
Image
of
Los
Alamos
Fires

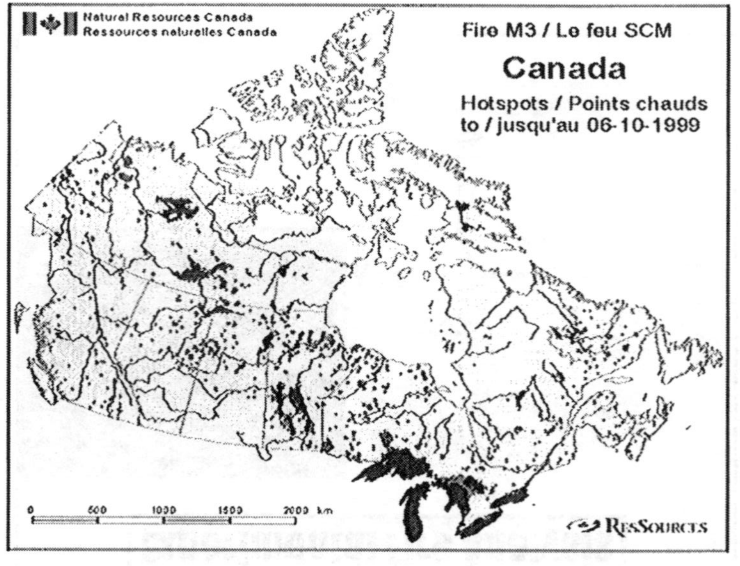

Natural Resources Canada
Ressources naturelles Canada
Fire M3 / Le feu SCM
Canada
Hotspots / Points chauds
to / jusqu'au 06-10-1999
0   500   1000   1500   2000 km
RESSOURCES

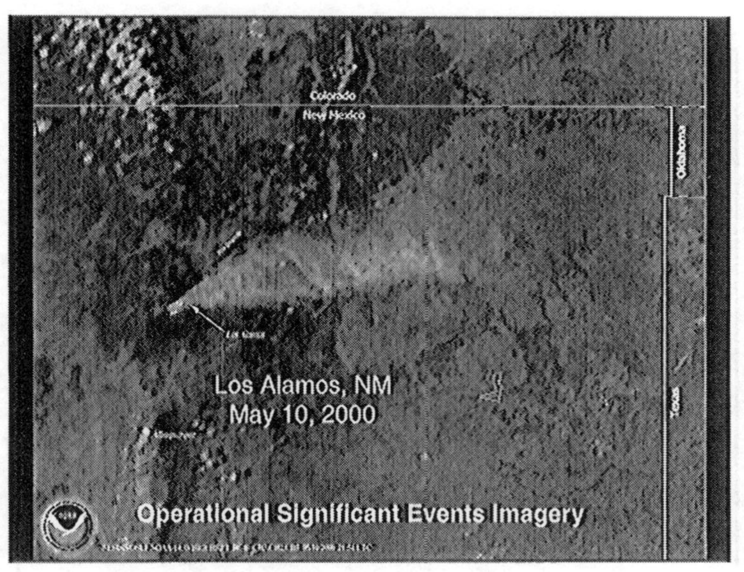

Colorado
New Mexico
Oklahoma
Texas
Los Alamos, NM
May 10, 2000
Operational Significant Events Imagery

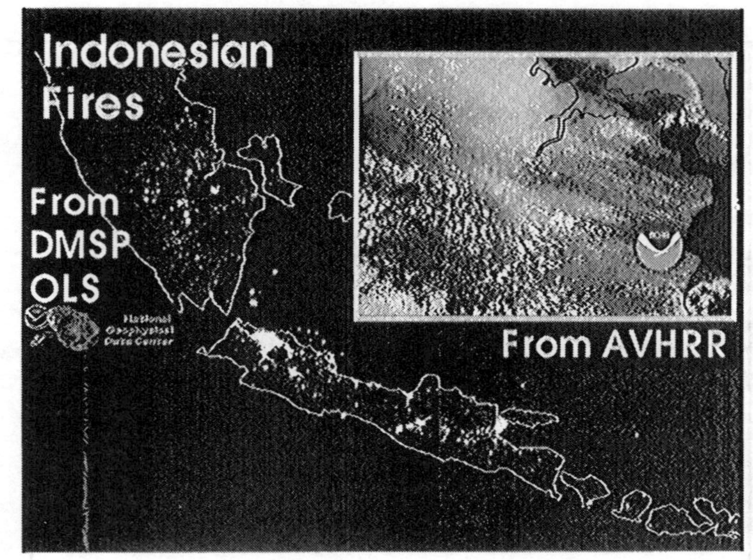

Indonesian
Fires
From
DMSP
OLS
National
Geophysical
Data Center
From AVHRR

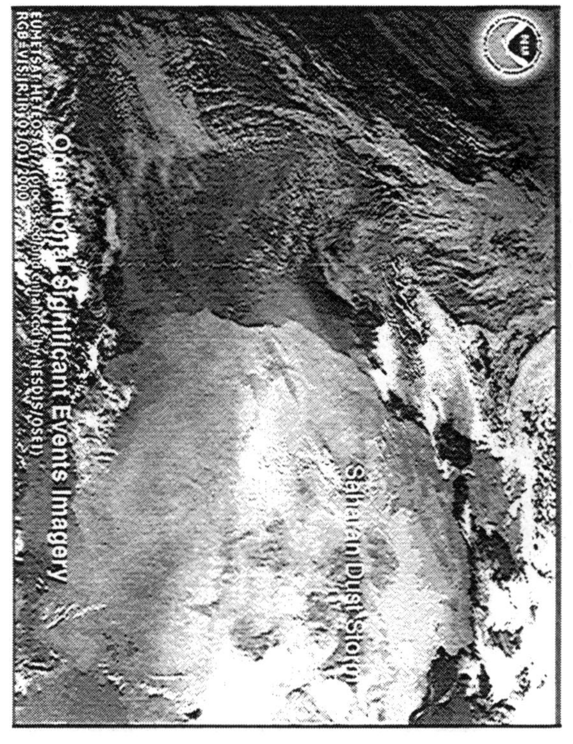

Operational Significant Events Imagery
Saharan Dust Storm

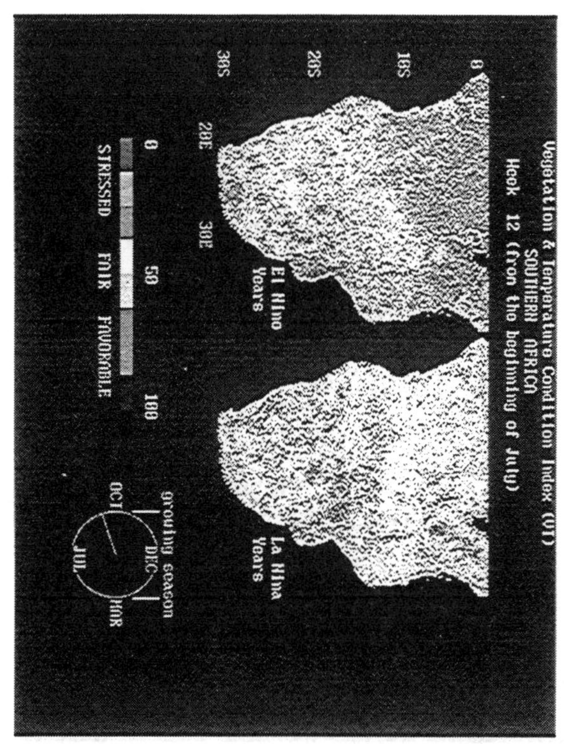

Vegetation & Temperature Condition Index (VT)
SOUTHERN AFRICA
Week 12 (from the beginning of July)
El Nino Years
La Nina Years
STRESSED   FAIR   FAVORABLE
growing season

SARSAT Rescues (1982 to 12/99):   11,100
1999 U.S. Rescues: 353

Commercial Remote Sensing Licensing
www.licensing.noaa.gov
ASTRO VISION
EARTH-WATCH
ORBIMAGE
SPACE IMAGING
BOEING
MARCONI
RDL

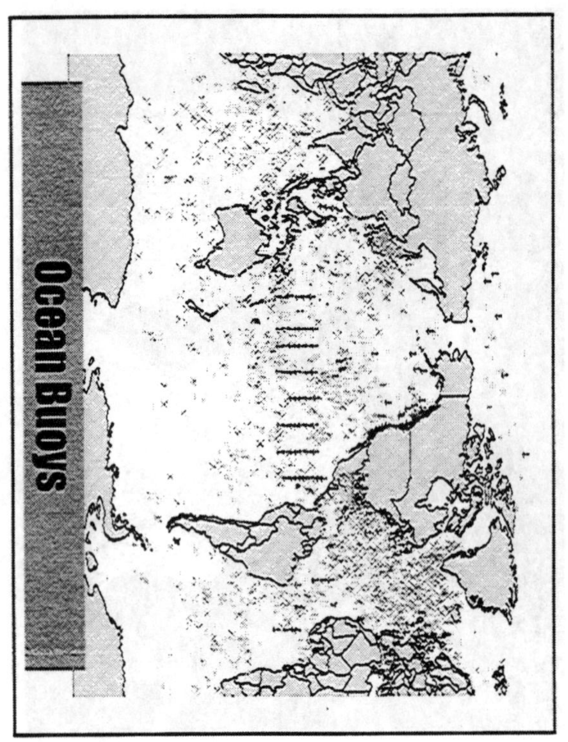

Ocean Buoys

Toga-Tao Drifting Buoys

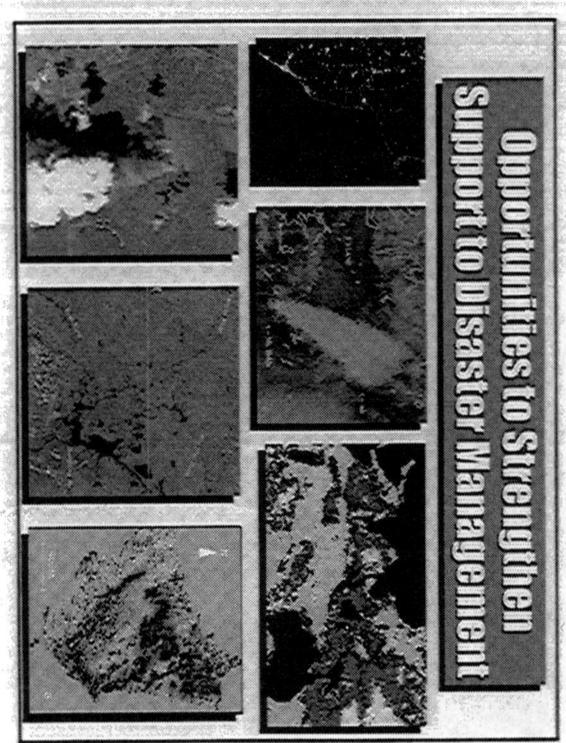

Opportunities to Strengthen
Support to Disaster Management

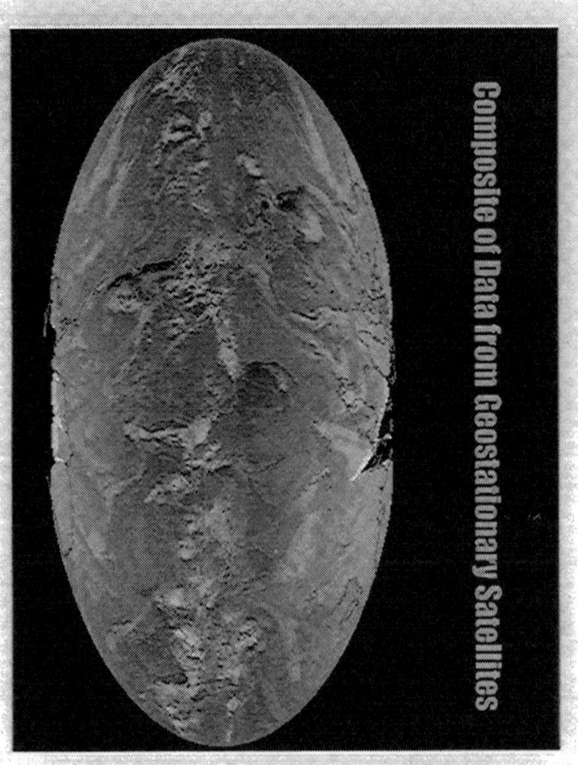

Composite of Data from Geostationary Satellites

A DEMONSTRATION OF THE USE OF SATELLITE TECHNIQUES
FOR EMERGENCY MANAGEMENT:
SPACE TECHNOLOGY FIGHTS FOREST FIRES IN CANADA THIS
SUMMER WITH REMSAT*

Steve Newton
Fire Manager, Lillooet Zone,
British Columbia Forest Service,
Bag 700,
Lillooet, British Columbia, V0K 1V0
steve.newton@gems4.gov.bc.ca

European Space Agency (ESA) contact:
Emmanuel Rammos
REMSAT Project Manager, D/APP-TNB
ESTEC, PO Box 299, 2200AG, Noordwijk, The Netherlands
erammos@estec.esa.nl

Existing and newly emerging space-based technologies (positioning, communications and Earth observation) could satisfy many of today's emergency management requirements, but a gap seems to exist currently between these technologies and their operational use.

THE EUROPEAN SPACE AGENCY (ESA) DIRECTOR OF APPLICATION PROGRAMMES (D/APP) REMSAT ACTIVITY

The primary objective was to demonstrate the use of real-time satellite communications, localisation, Earth observation and meteorological services in emergency situations using existing technologies via a pilot demonstration involving end-users. The work included:

- system architecture and optimisation
- integration and testing of a REMSAT system
- real-size simulations and evaluation
- a pilot demonstration in a real emergency
- direct involvement of end-users

Emphasis was on operational and financial sustainability of the system.

---

* *This paper was presented at the United Nations/Chile/European Space Agency Workshop on the "Use of Space Technology in Disaster Management", held from 13 to 17 November 2000 in La Serena, Chile, and does not necessarily reflect the views of the United Nations.*

The ESA contract was with MacDonald Dettwiler and Associates Ltd. (British Columbia, Canada) and the British Columbia Forest Service (BCFS) Protection Program. The BCFS Protection Program is a provincial government agency responsible for:

- fighting an average of 2,500 to 3,000 forest fires in an area of over 1 million square kilometres, an activity which contributes approximately $15 billion of annual economic activity
- providing logistics support and aviation management to the province in the event of a large natural disaster

The Protection Program's organizational structure consists of six regionally located fire centres and one headquarters. The headquarters provides policy and procedural direction as well as financial management. The fire centres manage regional administrative and operational fire suppression activities.

REMSAT is seen to provide a missing link in BCFS fire suppression activities. It is based on the Incident Command System (ICS) organisational structure. ICS is used by emergency response organisations throughout North America and has been adopted by many other countries as a standard system for managing all types of emergency operations involving multiple agencies. ICS is based on the premise that all activities in any emergency situation can be managed under one of the following functions:

- Incident Command
- Operations
- Planning
- Logistics
- Finance

Disasters often impact large or remote geographic areas where:

- Communications infrastructure is eliminated, damaged or overloaded
- On-scene personnel have little or out-dated geo-spatial information
- Managers need intelligence about the scope of problem and the required response level
- It may be difficult to plan response and communicate plans to on-scene staff

Currently, the technology and communications capability available to the on-site commander is often limited to line-of-sight VHF and UHF radio. The existing system is very expensive to maintain as it gets older. In 2004, Canada will be doubling the amount of bands within an assigned set of FM frequencies to accommodate increased demand for radio communications. In 2009, it will double again. Currently, technology is not commercially available to address the 2009 scenario. Satellites are a viable option.

Spatial information and communication is limited to verbal descriptions, sketches, and annotations to paper products characterised by slow delivery of information to and from the field. It can sometimes take three or four days to get good map products to the field.

The REMSAT concept is compatible with generic needs of other types of emergencies such as:
- earthquakes, floods, tsunamis and landslides
- exceptionally heavy winter conditions, or
- emergencies related to hazardous materials

REMSAT OBJECTIVES

To provide:
- communications
- position determination
- Earth observation services

This information meets the following emergency management needs:
- providing real-time communications between headquarters, command posts and remote field personnel
- providing independence from terrestrial systems
- monitoring field crew position and status
- supporting response planning through detailed up-to-date digital map information

REMSAT DESIGN

A multiple satellite system is baselined for the first demo:
- Localisation via the Global Positioning System (GPS)
- Messaging via the ORBCOMM low-Earth orbit (LEO) satellite system
- 128 bps data rates, voice and video services via the ANIK geosynchronous orbit (GEO) satellite system. (2 voice, 1 fax and 1 data lines are dedicated)

REMSAT can be adaptable to alternative satellite systems selection depending on geographical area of interest. The basic three-level structure is:
- Control Centre Terminal (CCT) for central command in a safe, permanent location
  - long term data collection and storage
  - data integration and packaging
  - global asset allocation and monitoring
- Intermediate Mobile Terminal (IMT), a transportable command center on-site
  - Field operations management
  - situation assessment
  - activity planning
  - resource allocation and tracking
- Hand Held User Terminal (HUT) for unit-level data/voice communications
  - location, status
  - events, activities, reporting

Management of fires in the field requires specific information:
- Location
- Hazards
- Values at risk
- Weather
- Terrain
- Access to the site
- Water sources
- Land use/land cover/land ownership
- Ease of request and timely delivery

Field planning activities with REMSAT use:
- Projector to dry erase white board for personnel interaction and annotation of digital map
- Multiple viewers can observe projected images
- Digital documention of features after decisions and actions
- Multiple Geographic Information System (GIS) overlays
- Freeze/print/distribute map products to field in timely manner

The REMSAT system provides:
- Satellite communication
- Desktop and a portable personal computer
- Wireless Local Area Networks (LAN)
- Full Internet capabilities and access to internal network information systems
- Printer
- Telephone/fax
- Base radio transceiver
- Uninterruptable and portable power source
- Air conditioning

HUT objectives:
- Provide smallest possible unit for human transport
- Critical function is position determination & transmission
- Avoid line-of-sight issues
- Two-way text messaging

The REMSAT pilot project used ORBCOMM text messaging facility as well as GPS

HUT Capabilities:
- Crew localization
  - GPS positioning
  - Waypoint marking

- Messaging:
  - Canned and free text

* Fire Weather Advisory
* Spot weather report
* Resource requests
* Initial Fire Reports

PROJECT KEY EVENTS SCHEDULE

* Proof of Concept demonstration, Abbotsford British Columbia, September, 1999
* System Acceptance Test passed beginning of March, 2000
* Demonstration #1: Squamish, British Columbia, May, 2000
* Demonstration #2: Lillooet District, British Columbia, May, 2000 (Demonstrations 1 and 2 are multi-agency "wildland urban interface" exercises)
* First operational deployment of REMSAT on 11 August 2000

OPERATIONAL DEPLOYMENT

REMSAT, installed by BCFS staff on-site, was operational within two hours and 40 minutes of arrival. The deployment was considered as highly successful by BCFS because:

* Established stable and robust communications in earliest stages of incident
* Access to all internal information systems with no noticeable disruptions or delays
* Incident historical documentation and archiving began as soon as the power was turned on
* Delivery of high quality mapping to field crews was almost immediate
* Support provided for a second major fire from one location
* Provincial fire behavior specialist relocated to fire camp and was able to maintain support for two fires as well as the rest of the province

Following a full evaluation of the results, the project will end with the final presentation, which is foreseen by the end of October.

GENERAL RECOMMENDATIONS

* Further development of REMSAT capabilities with other agencies: mutual contributions and standardized approach to emergency management
* Further emergency scenarios in more complex environments
* Further geographic testing, also with different commercial satellite vendors
* Test availability of readily available commercial Earth observation products
* Incorporate other "domain-specific" software for more versatility
* Eliminate forest fire bias and become more generic
* Continue to explore HUT technologies, such as smaller dishes and dual-mode wireless phones

"Un programa de transferencia para incorporar el uso de la tecnología espacial en los planes de protección civil en el Perú"

Ing. Ricardo Coloma
Comisión Nacional de Investigación y Desarrollo Aeroespacial
CONIDA
Felipe Villarán 1069, Lima 27
Telefax 511-4419081
rcoloma@conida.gob.pe
http://www.conida.gob.pe

RESUMEN

En la búsqueda de reducir el impacto de los desastres naturales en la economía nacional y a la población, el programa de percepción remota de CONIDA, elaboró una estrategia tendiente al establecimiento de la mejor metodología de prevención y monitoreo de zonas sensibles así como los mecanismos de transferencia de dicha metodología.

Haciendo uso permanente de las capacidades establecidas en el Centro de Estudios Espaciales de la Institución, se han programado cursos mensuales de diseminación de la tecnología de percepción remota y sistemas de información geográfica, cursos que vienen permitiendo incrementar sustancialmente el número de usuarios nacionales.

La utilización de los laboratorios de procesamiento de imágenes de la Comisión, diferentes metodologías son experimentadas y puestas a disposición como material educativo para los estudiantes de los cursos.

Finalmente permanentes convenios suscritos con Universidades Estatales de todo el territorio nacional, el programa de percepción remota implementará nuevos mecanismos de transferencia de tecnología.

En coordinación con el Instituto Nacional de Defensa Civil- INDECI y la red de instituciones que han suscrito acuerdos con CONIDA, la tecnología espacial será promovida hacia su utilización en mecanismos de protección civil.

** This paper was presented at the United Nations/Chile/European Space Agency Workshop on the "Use of Space Technology in Disaster Management", held from 13 to 17 November 2000 in La Serena, Chile, and does not necessarily reflect the views of the United Nations.*

"Un programa de transferencia para incorporar el uso de la tecnología espacial en los planes de protección civil en el Perú"

1.- INTRODUCCIÓN

En la búsqueda de reducir el impacto de los desastres naturales en la economía nacional y en la población, el programa de percepción remota de CONIDA, a elaborado una estrategia tendiente al establecimiento de la mejor metodología de prevención y monitoreo de zonas sensibles así como los mecanismos de transferencia de dicha metodología.

Haciendo uso permanente de las capacidades establecidas en el Centro de Estudios Espaciales de la Institución, se han programado cursos mensuales de diseminación de la tecnología de percepción remota y sistemas de información geográfica, cursos que vienen permitiendo incrementar sustancialmente el número de usuarios nacionales.

La utilización de los laboratorios de procesamiento de imágenes de la Comisión, motivó diferentes metodologías a ser experimentadas y puestas a disposición como material educativo para los estudiantes de los cursos.

Finalmente la suscripción de convenios con Universidades Estatales de todo el territorio nacional, permite al programa de percepción remota implementar nuevos mecanismos de transferencia de tecnología.

En coordinación con el Instituto Nacional de Defensa Civil- INDECI y la red de instituciones que han suscrito acuerdos con CONIDA, la tecnología espacial será promovida hacia su utilización como mecanismo de protección civil.

El abaratamiento de los mecanismos de transferencia de información, como es el caso de los formatos FTP a través del canal de INTERNET facilitará a los suscriptores de la información contar con alertas completamente documentadas.

2.- TIPOS DE DESASTRES NATURALES COMUNES EN EL PERU

Debido a las características Geo-dinámicas del territorio nacional, el mayor riesgo existente son los sismos asociados principalmente al comportamiento de la placa de Nazca. Por ello el permanente monitoreo de la sismisidad local es un factor importante.

En el Perú existen varios volcanes semi-activos pero en los últimos años no se les puede atribuir desastres ocasionados directamente a ellos.

Otro fenómeno importante en los últimos años a sido el Niño que por su gran amplitud afecto a todo el territorio nacional produciendo grandes pérdidas de infraestructura, producción y vidas humanas.

En ambos casos las zonas afectadas del territorio nacional son muy amplias y su monitoreo suele realizarse posteriormente los hechos con la finalidad de evaluar el impacto.

Un fenómeno común y estacional son las avalanchas o huaycos que suelen ocurrir en los meses del verano productos de intensas lluvias que son fáciles de predecir. Estos huaycos suelen cortar las carreteras y vías férreas afectando el traslado de productos y personas al y desde el interior del país.

El desborde de ríos es un fenómeno que suele tener importancia en las zonas de la selva, donde por la facilidad de ocupación las poblaciones suelen ser ribereñas y son profundamente afectadas cuando el torrente de los ríos supera las zonas pobladas, arrasando también con los cultivos.

Casi todos los años se producen en la serranía grandes heladas que destruyen los cultivos y sobre las cuales poco se ha desarrollado.

3. - PROYECTOS PILOTO DESARROLLADOS POR CONIDA

Con la finalidad de afinar las metodologías, el programa de percepción remota ha venido ejecutando por cuenta propia diversos proyectos sobre variedad de escenarios del territorio nacional, que permitan analizar todos los procedimientos de un proceso de monitoreo o de evaluación del impacto de un fenómeno natural de importancia.

3.1 Proyector Glaciar

En los últimos años se habla sobre el calentamiento global y este proyecto pudo constatar en la mayor zona de glaciares del país, la retracción de los mismos y la menor área cubierta de hielos en la Cordillera Blanca.

Mediante imágenes satelitales se logró clasificar hasta siete variedades de hielos y cinco variedades de lagunas, así mismo

el número de hectáreas cubiertas de hielo.

3.2 Proyecto Lima-Pisco

La permanente contaminación oceánica y el impacto sobre las zonas urbanas fueron detectados por este proyecto que cubrió una extensión de 300 Km y en el cual se utilizaron imágenes ópticas y de radar.

3.3 Proyecto Minsienor

El fuerte impacto de la actividad minera sobre las áreas influenciadas, fue detectado por imágenes de satélite con

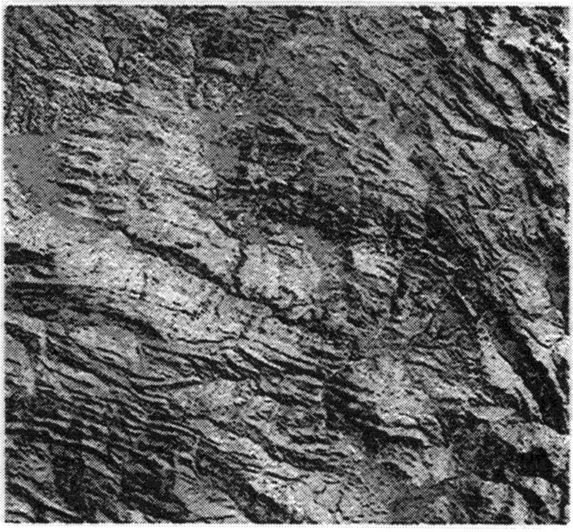

gran exactitud. La cuantificación de las zonas dañadas e irrecuperables se verificaron con trabajos de campo.

3.4 Proyecto Sabancaya

La simulación de una posible erupción del volcán Sabancaya y las posibles corrientes de lava en los alrededores fue simulada a partir de imágenes de satélite con gran exactitud.

3.5 Proyecto Ciudad Iquitos

El constante desplazamiento del Río Amazonas ha afectado a la ciudad de Iquitos por años, la utilización de imágenes de radar y ópticas ha permitido

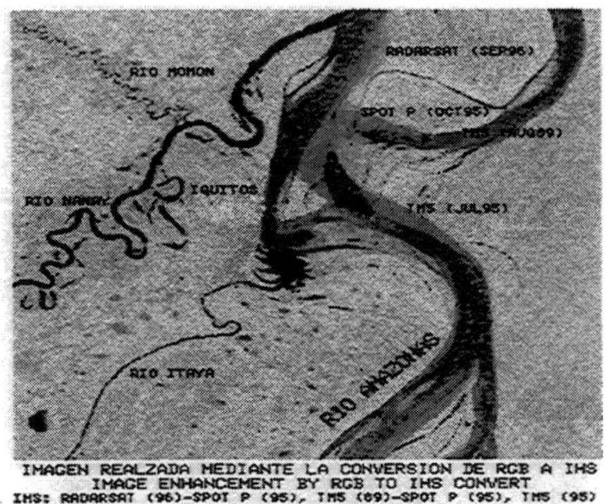

cuantificar el real desplazamiento del río y su impacto sobre la ciudad.

3.6 Proyecto Mantaro

La escasez de recursos hídricos para la Central del Mantaro fueron repotenciadas por la evaluación de un sin número de nuevas lagunas en la vertiente del río Mantaro. Simulaciones y 3D se realizaron con la finalidad de evaluar el impacto de un mayor caudal producido por el rompimiento de las lagunas ante un exceso de lluvias.

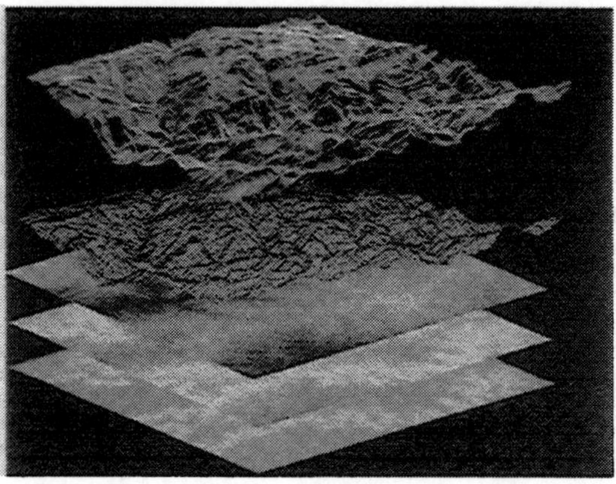

4.-RED DE INFORMACIÓN DE TECNOLOGÍA EN PERCEPCIÓN REMOTA

Mediante la suscripción de múltiples acuerdos de cooperación, CONIDA a través del Programa de Percepción Remota viene implementado una red que se constituirá en los futuros usuarios de los datos que oportunamente proporcionará el minisatélite CONIDASAT-01.

En la estructura de la red se encuentran desde Universidades hasta empresas privadas.

Universidades Estatales

- Universidad Nacional Mayor de San Marcos

- Universidad Nacional Agraria la Molina
- Universidad Nacional de Ingeniería

Próximas suscripciones en trámite:

- Universidad Nacional José Faustino Sánchez Carrión
- Universidad Nacional San Antonio Abad
- Universidad Nacional del Santa
- Universidad Nacional Daniel Alcides Carrión
- Universidad Nacional de Piura
- Universidad Nacional Agraria de la Selva
- Universidad Nacional de Huancavelica
- Universidad Nacional San Agustín de Arequipa
- Universidad Nacional del Altiplano
- Universidad Nacional Jorge Basadre de Tacna

A nivel de Instituciones Públicas se tiene acuerdos con:

- Instituto Geofísico del Perú
- Instituto de Defensa Civil
- Consejo Nacional de Ciencia y Tecnología
- Servicio Nacional de Meteorología e Hidrología

En trámite:

- Instituto del Mar del Perú
- Instituto Geológico Minero y Metalúrgico
- Instituto Nacional de Recursos Naturales
- Instituto Geográfico Nacional

- Instituto Nacional de Telecomunicaciones

Otros en trámite:

- Instituto de la Amazonía Peruana Consejo Nacional del Medio Ambiente
- Registro Público de Minería
- Instituto Metropolitano de Planificación

5.-CONVENIO CONIDA-INDECI

Mediante la utilización de la actual infraestructura de CONIDA y de su Centro de Estudios espaciales se ha brindado cursos de introducción y especialización a mas de 1200 especialistas sobre procesamiento de imágenes, sistemas de información geográfica y uso de GPS, lo que garantiza el número de usuarios especializados a nivel nacional.

Así mismo en el marco del convenio suscrito entre CONIDA-INDECI, se dio cabida recientemente en sus instalaciones al curso que organizó el INDECI para funcionarios de todas sus regiones en la utilización del SIG como herramienta de planeamiento de las acciones a tomar en caso de desastres naturales.

En la búsqueda de una relación mas estrecha entre las instituciones se tiene en desarrollo la realización de un nuevo proyecto piloto en el cual ambas instituciones encuentren solución a sus necesidades y en cual se utilice el máximo de la tecnología espacial de percepción remota que CONIDA ha demostrado dominar.

A través de la red a implementarse con las Universidades y en coordinación con el INDECI, se implementará un mecanismo de educación a distancia con la finalidad que nuevos participantes de todo el territorio nacional tenga la capacidad de manejo de la tecnología y sea de apoyo a las operaciones regionales de Defensa Civil.
La reciente participación de profesores de Universidades de provincia en el Simposio Nacional que organiza todos los años CONIDA son una muestra de ello.

6.- PROYECTO PILOTO

Lima es una ciudad que tiene mas de ocho millones de habitantes para la cual no se ha desarrollado un SIG en alta resolución orientado a apoyar la reacción a un gran sismo.
La necesidad de ubicar Hospitales, Clínicas Privadas, Farmacias, Vías rápidas, Colegios, Grandes Edificios, Cuarteles, Asilos de ancianos, Cementerios, Líneas de Alta Tensión, Líneas de Agua y Desagüe, etc. Son una necesidad importante para el planeamiento de la reacción en caso de un gran sismo y particularmente para difusión entre la población para que sepa como actuar de presentarse este gran fenómeno telúrico.
El programa de percepción remota de CONIDA, tiene interés en desarrollo de modelos dinámicos que permitan visualizar los efectos que podrían suscitarse en caso de un gran sismo.

7.- CONCLUSIÓN

La tecnología espacial y la utilización de los Sistemas de Información Geográfica han demostrado ampliamente la capacidad de visualización de cambios siendo especialmente útil para la prevención de desastres naturales, el planeamiento de acciones preventivas y de reacción en caso de suceder.
La vinculación con la mayor cantidad de centros Universitarios de todo el país permitirá formar y especializar al mayor número de profesionales que serán el mejor soporte a las oficinas regionales del INDECI a través del masivo número de alumnos entrenados.

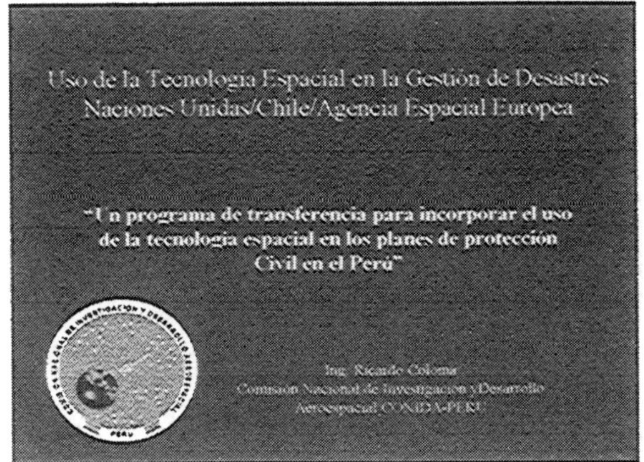

Uso de la Tecnología Espacial en la Gestión de Desastres
Naciones Unidas/Chile/Agencia Espacial Europea

"Un programa de transferencia para incorporar el uso
de la tecnología espacial en los planes de protección
Civil en el Perú"

Ing. Ricardo Coloma
Comisión Nacional de Investigación y Desarrollo
Aeroespacial CONIDA-PERU

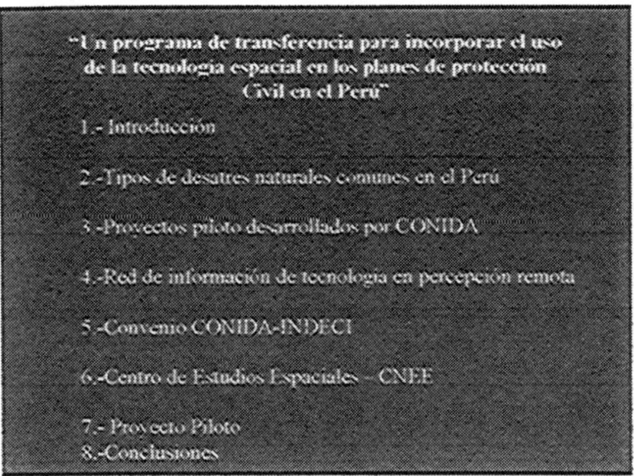

"Un programa de transferencia para incorporar el uso
de la tecnología espacial en los planes de protección
Civil en el Perú"

1.- Introducción

2.- Tipos de desatres naturales comunes en el Perú

3.- Proyectos piloto desarrollados por CONIDA

4.- Red de información de tecnología en percepción remota

5.- Convenio CONIDA-INDECI

6.- Centro de Estudios Espaciales – CNEE

7.- Proyecto Piloto
8.- Conclusiones

"Un programa de transferencia para incorporar el uso
de la tecnología espacial en los planes de protección
Civil en el Perú"

TECNOLOGIA

CONIDA

USUARIOS

Fenómeno del Niño

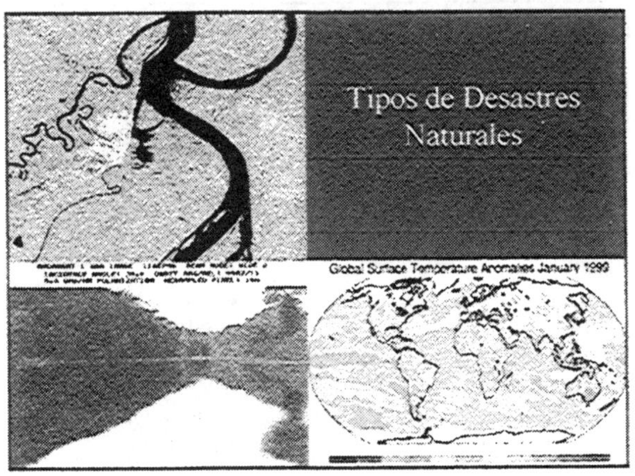

Tipos de Desastres
Naturales

Global Surface Temperature Anomalies January 1999

Tipos de Desastres
Naturales

Sismos

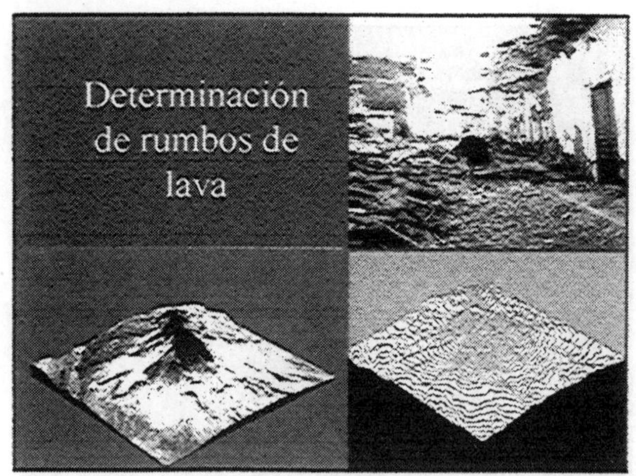

Determinación
de rumbos de
lava

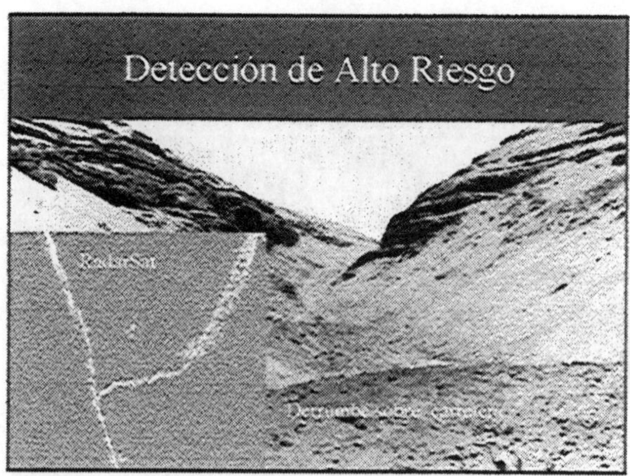

Detección de Alto Riesgo
RadarSat

Proyectos Piloto Desarrollados
por CONIDA

SIG – Zonificación de riesgo

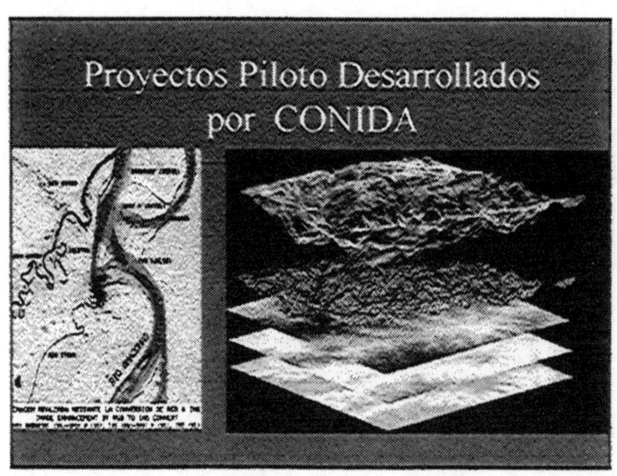

Proyectos Piloto Desarrollados
por CONIDA

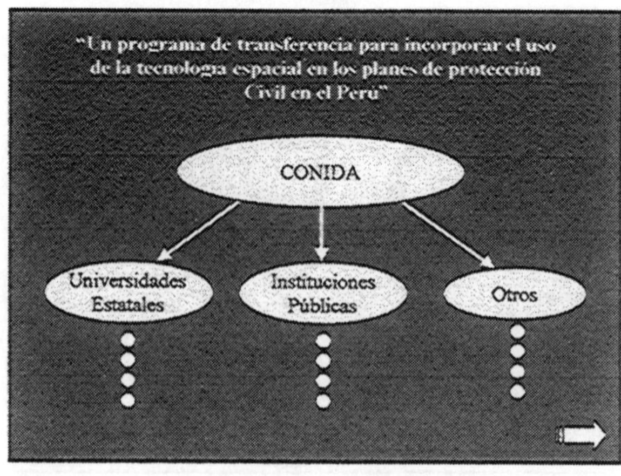

"Un programa de transferencia para incorporar el uso
de la tecnología espacial en los planes de protección
Civil en el Perú"
CONIDA
Universidades
Estatales
Instituciones
Públicas
Otros

Universidades Estatales

Universidad Nacional Mayor de San Marcos
Universidad Nacional Agraria La Molina
Universidad Nacional de Ingeniería
Universidad Nacional José Faustino Sánches Carrión(*)
Universidad Nacional San Antonio Abad del Cuzco(*)
Universidad Nacional del Santa(*)
Universidad Nacional Daniel Alcides Carrión(*)
Universidad Nacional de Piura(*)
Universidad Nacional Agraria de la Selva(*)
Universidad Nacional de Huancavelica(*)
Universidad Nacional San Agustín de Arequipa(*)
Universidad Nacional del Altiplano(*)
Universidad Nacional Jorge Basadre de Tacna(*)

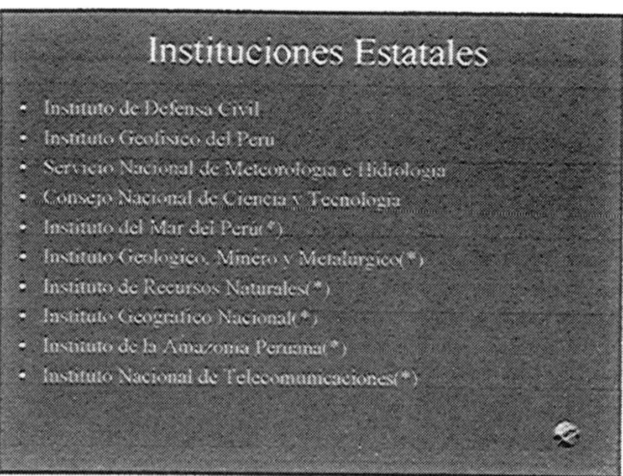

Instituciones Estatales

Instituto de Defensa Civil
Instituto Geofísico del Perú
Servicio Nacional de Meteorología e Hidrología
Consejo Nacional de Ciencia y Tecnología
Instituto del Mar del Perú(*)
Instituto Geológico, Minero y Metalúrgico(*)
Instituto de Recursos Naturales(*)
Instituto Geográfico Nacional(*)
Instituto de la Amazonía Peruana(*)
Instituto Nacional de Telecomunicaciones(*)

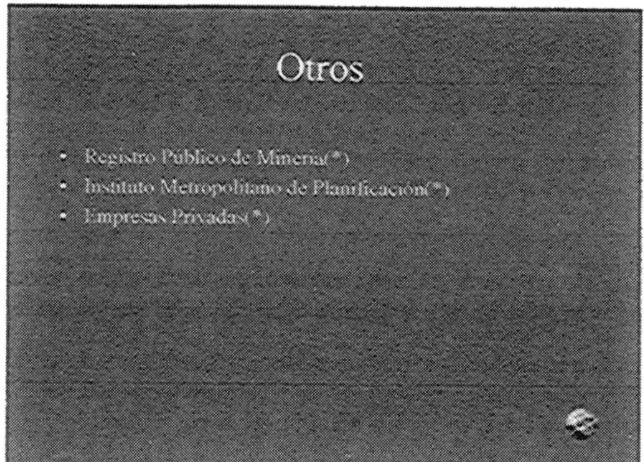

Otros

Registro Público de Minería(*)
Instituto Metropolitano de Planificación(*)
Empresas Privadas(*)

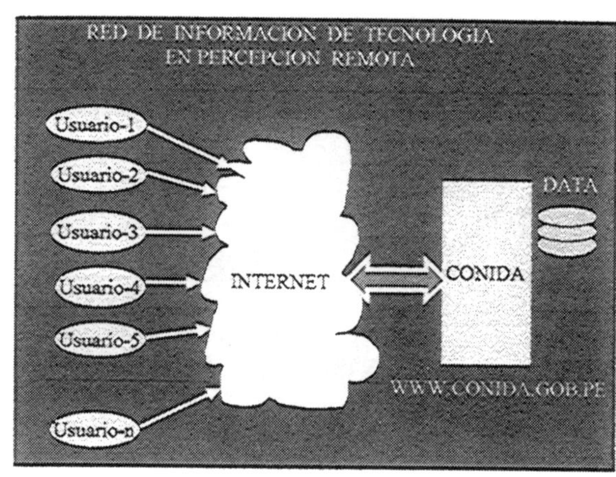

RED DE INFORMACION DE TECNOLOGIA
EN PERCEPCION REMOTA

Usuario-1
Usuario-2
Usuario-3
Usuario-4
Usuario-5
Usuario-n
INTERNET
CONIDA
DATA
WWW.CONIDA.GOB.PE

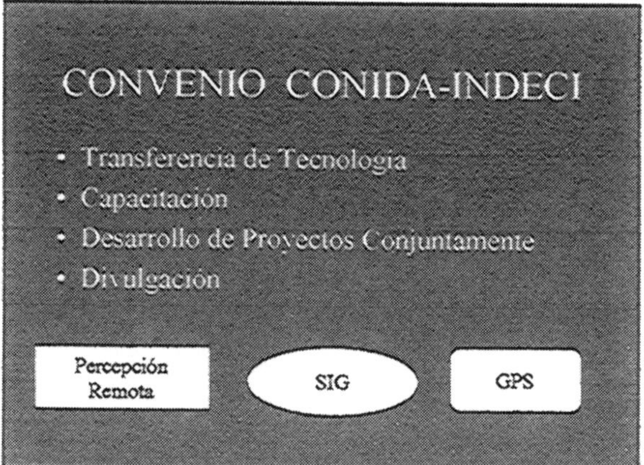

CONVENIO CONIDA-INDECI

Transferencia de Tecnología
Capacitación
Desarrollo de Proyectos Conjuntamente
Divulgación

Percepción Remota
SIG
GPS

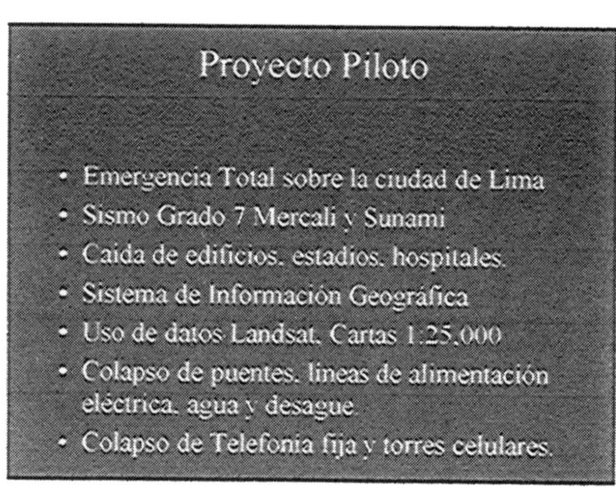

Proyecto Piloto

Emergencia Total sobre la ciudad de Lima
Sismo Grado 7 Mercali y Sunami
Caida de edificios, estadios, hospitales.
Sistema de Información Geográfica
Uso de datos Landsat, Cartas 1:25.000
Colapso de puentes, lineas de alimentación eléctrica, agua y desague.
Colapso de Telefonía fija y torres celulares.

CENTRO DE ESTUDIOS
ESPACIALES

- Cursos Mensuales de percepción remota, sistemas de información geográfica y de posicionamiento global.
- Programa de Maestría con títulos a nombre de la Nación
- Cursos a distancia (*)
- Simposio Nacional de Ciencia y Tecnología Espacial